Advance Praise for

Somebody Hold Me is the textbook for the touch we need. Jordan clarifies the lines of sexuality, health, and well-being, and shows a path for healthy human touch that might save us from our isolated, screen-dominated future.
— JOHN HALCYON STYN, *Hug Nation*

Somebody Hold Me is a smart, practical, and empowering guide for anyone who could benefit from more nurturing touch in their 21st century lives, which, Jordan helped me realize, is pretty much all of us.
— KIM CORBIN, *founder of iSkip.com*

If you're not getting touched, then this is the book for you. Somebody Hold Me is the how-to manual about touch that the world has been waiting for – without knowing it. Warning! Following these instructions could positively change your health, uplift your attitude, and brighten your world.
— BELLA LAVEY, *Fetish Girl: A Memoir of Sex, Domination, and Motherhood*

Funny, engaging, and down to earth, Somebody Hold Me helps readers defeat touch deprivation with an often-untapped resource: the loving support of friends.
— CHRISTINE HOFF KRAEMER, PhD, author of *Eros and Touch from a Pagan Perspective* and editor of *Pagan Consent Culture*

Somebody Hold Me is the first book of its kind. Touch is invaluable, accessible, within reach, and worth seeking!
— JANET TREVINO, ReTouch Rehabilitative Touch Therapy

Somebody Hold Me

The Single Person's Guide
To Nurturing Human Touch

Somebody Hold Me

The Single Person's Guide To Nurturing Human Touch

Epiphany Jordan

KARUNA SESSIONS
Austin, Texas

[Paperback]
ISBN 13: 978-1-7328792-0-1

[E-Book]
ISBN 13: 978-1-7328792-1-8

1. Friendships 2. Cultural Studies
3. Healthy Living I. Book Title

Library of Congress Control Number: 2018913332

Author photo credit: Vanessa Filkins
Book cover art: "Held" by Eliza Bundledee
All rights reserved.

Ordering Information:
Inquire for wholesale or group pricing.

Printed in the U.S.A.

Significant Otter Publishing
Austin, TX
www.nurturinghumantouch.com
info@nurturinghumantouch.com

Table of Contents

Part 2: The Step-by-Step Guide

Part 3: The Real World

For Mac and Cheese,
who gives the best hugs.

"The most natural way for
human beings to calm themselves
is by clinging to another person."

Bessel van der Kolk
The Body Keeps the Score

Introduction

I T WAS 3:30 a.m., and I was awake. The wee hours of the night and I had become well-acquainted in the past year, but that night was filled with tears. Sobs escaped my mouth as I curled into a ball around a pillow. Grief galloped back into my mind and my heart. It was the first of many early mornings I would spend crying.

The day before, my partner and I had decided to part ways. He has one story, I have another. As with many breakups, there were compelling reasons to stay, and equally compelling reasons to go. It took me nearly a year to decide to end it.

We had met five years earlier through Craigslist Casual Encounters (that's how hookups happened before you could swipe right). I wasn't expecting much, but this hookup was good. Really good. Soon we were hooking up two to three times a week, and after six months discovered we had fallen in love.

As we got to know each other, I learned more about his relationship with his body. The "meat cart," as he referred to it, was the worst part of being alive. He would have paid a very large sum of money to upload his consciousness to a server somewhere if it meant that he would no longer have to inhabit the horrible thing.

Despite his hatred of the meat cart, of all the things I miss about our relationship, it's his body that I miss the most. For the first time, I had a relationship that was touch-centric, and I felt healthy and happy. I laughed a lot, and became a warmer, more open version of myself. We rarely slept in the same bed – I'm a light sleeper and he refused to use his CPAP machine – but we hugged and snuggled and touched frequently. I loved the way his arms fit around my waist and his hands rested on the small of my back. Even though he smoked cigarettes, the smell of his body was a symphony of sweetness in my nostrils. Some of our best times were spent in the bathtub, with loads of fragrant bubbles courtesy of LUSH, me lying back on his chest, his fingers lightly stroking my shoulder. I would run my fingers along the hair and scars on his arms, listen to his heartbeat, and know I was safe and cared for.

The body he hated so much gave me pleasure, comfort, and joy. Sometimes I can still feel the imprint that his body has left on mine, and I wonder how long it will be until it fades.

After the breakup, my friends hugged me when I cried. While my grief has been deep – I have never loved anyone as much as I loved him – I have been supported. After a few months, I got back to a rhythm of solo living, and other people's crises dominated my social circles.

In the year since the breakup, I've come to realize that I may never have a sexual or romantic relationship again. I am a pragmatist, and know that the pool of available partners is much smaller now that I'm in my 50s. The last two years of my relationship left me emotionally, mentally, and financially drained, and I am wary of merging my life again. I enjoy living alone, and many men I came across were interested in having a girlfriend or wife live with them and take care of them. I have goals and dreams I want to focus on, and "falling in love" is not high on the list.

Yet trying to find a casual lover was an exhausting and annoying process. In my experience, this sort of relationship takes a great deal of respect, maturity, and excellent communication skills, and this is exactly the sort of intimacy hook-up culture avoids. One-night stands hold no interest for me after five years of hot, heart-connected sex. I am a sapiosexual – I'm attracted to intelligence – and, for me, sexual chemistry takes longer to build than one conversation over a meal or a drink.

The men I chatted with lost interest when they discovered that sex wasn't happening within hours of our first email exchange; they wanted the benefits without the friends. After a few months, I deleted all my dating profiles. (Sadly, bisexuality is not an option for me, and I don't have time for the intricacies of polyamory.)

As I contemplate the next 30 or so years of my life, I know I will have many of my needs met: companionship, collaboration, conversation, adventures, and fun will be plentiful. I have long-term, supportive friendships that feed me emotionally, and people I can call when I need a ride home from the doctor. I can masturbate myself into mind-blowing, endorphin-filled, multiple orgasms. But somebody to snuggle with? Under our current model of not separating touch from romantic relationships and sex, I'm S.O.L.

There has to be a better option for me and the millions of single people who are yearning for somebody to hold them.

Lack of touch is an epidemic, and nurturing human touch is the cure. For the past five years, I've seen firsthand the benefits of touch with my business, Karuna Sessions. A Karuna Session is a ritual of human experience that culminates in the client being held between two practitioners.

Yes, it is as blissful as it sounds.

I've watched men in their 60s whose frowns soften into gentle smiles after being held by us, and I've heard the relieved sobs of new mothers who are receiving some tender touch, after constantly giving it. I've felt tense bodies relax deeply as they reset to the feeling of peace and safety they experienced in their mothers' arms.

But if nurturing human touch is so crucial, why is it in such short supply? How do we bridge the gap between one human and another? How did we go from sleeping in big piles of bodies in caves to apologizing when we bump into a stranger in the grocery store? Why have we created a web of shame, pride, fear, and embarrassment that keeps us physically separated from others?

This book was written to address those questions, and bridge that gap. It will help you navigate the increasingly fraught and fearful space between us. I have also created a framework that will allow you to share nurturing human touch relationships with the people you already love and trust: your friends.

Please note: This is not a guide about how to find or attract a partner, or how to have better sex. Millions of pixels have been sacrificed to that cause already, many of them eloquently. This guide will give you tools to improve your communication and negotiation skills, and get your touch needs met. While these things could help you in your quest to find a mate, that's not my focus. Still, there are nuggets o' wisdom aplenty in here for people who are in romantic or sexual relationships, and this information just might contribute to more harmonious partnerships.

Nurturing human touch is free, abundant, and organic, and giving and receiving it requires little training. It's a simple life hack with profound consequences. If you are willing to take a risk and connect with others, you will be able to enjoy better health and relationships.

Part I

The Importance of Touch

## CHAPTER 1

Who Is This Book For?

I F YOU'RE A person who isn't getting their touch needs met in the current "standard" way – through a romantic or sexual relationship – this book is for you.

If you've been told you're too intense, too old, too ugly, too fat, too crazy, too broke, too annoying, too weird, too depressed, or too disabled to have anyone fall in love with you, this book is for you.

If your inbox is filled with offers of sex, but you've discovered that none of your hookups want to snuggle afterwards, this book is for you.

If you've been friend-zoned more times than you've been on a date, this book is for you.

If you can't remember the last time someone other than your doctor or your hairdresser touched you, this book is for you.

If you're recovering from trauma, feel distrustful of others, and don't know if you can handle the emotional and physical intimacy of sex, this book is for you.

If you are recently divorced and don't miss your spouse leaving the lid off the toothpaste, but you sure as hell miss sleeping in the same bed, this book is for you.

If you wonder what all the fuss is about when people talk

about sex, but you enjoy tenderness, touch, and comfort, this book is for you.

If you have adult children now, and crave those times when your offspring's bodies snuggled into yours for warmth and comfort, this book is for you.

If you have a fantastic circle of friends already, yet don't interact with them other than the occasional hug, and you want more physical contact, this book is definitely for you.

Dating in the 21st century is a pain in the ass. I hear it from all my single friends. It sucks in different ways for men and women, but it sucks nonetheless. It's exponentially harder for those who don't identify as cisgender or heterosexual. "What's wrong with me that I can't find a partner?" is a common question. Nothing's wrong with you; it's hard for everybody.

The Missing Connection

Touch hunger – the craving for human contact – looks a bit different for every person. You may know you suffer from lack of touch; or perhaps you know that something is missing, but you don't realize what it is. You drum your fingers, tap your toes, touch your face, twirl your hair, but still....you're restless. Exercise, yoga, meditation, or dancing can get you out of your head for a bit, but it's challenging to stay focused and you often find yourself watching the clock.

Perhaps the missing connection is human contact via nurturing human touch.

Touch is complicated. Physiologically, it impacts our skin, heart, blood pressure, brain, nervous system, and muscles. Touch, or the lack thereof, also affects our thoughts, emotions, and worldview. Our attitudes toward touch are grounded in our culture and vary hugely across eras and regions. Touch flirts with issues of gender, human development, evolution, anthropology, sociology, sexuality, belonging, hierarchies, economics, communication, and personal preferences. It's intersectional, integrative, and increasingly important in a world where people feel isolated.

Proceed with Caution

Touching our friends is uncharted territory, and it requires interactions that don't have common social scripts. The idea may scare you, or make you nervous. If you use the system you will learn here, and work through the awkwardness, you can add nurturing human touch to your existing relationships in a thoughtful, consensual, fun way. You will learn how to create interactions where all y'all know exactly what's happening, what the desired outcome is, and that said outcome will be delightful for all participants.

"Reach out and touch someone" was a swell advertising jingle, but in this day and age, doing so could get your ass arrested. If you want to do this right, it will require being thoughtful. Please read the chapters in order, and take time to digest them. You will need to wrap your head around a whole bunch of different concepts before you start wrapping your arms around other people.

Group Hug!

A few years back, Karuna Sessions decided to offer cuddle parties (we called 'em Snuggle Salons) as one of our services. What I discovered is that people didn't need to learn how to hug – they already knew how to do that – but they did need to learn about boundaries and negotiating. They also needed to learn about the cultural sauce we're marinating in so they could challenge some of the commonly-held assumptions about touch and relationships.

The result? After two hours of face caressing, hugging, foot rubs, and snuggle sandwiches, a group of folks who had previously been strangers were smiling and grinning like they had taken some fine Molly and spent their evening rolling in a chill space at an event. The cool part? No teeth were clenched in the making of this experience. When we went around the circle at the end, participants used words like "happy," "peaceful," "content," "relaxed," and "satisfied."

Most people I know can use more "happy," "peaceful," "content," "relaxed," and "satisfied" in their lives. And let's face it: Most people no longer want to run around a party all night looking for some sketchy dude who goes by the name of Marshmellow so they can take a pill with iffy ingredients and spend the next few days feeling cracked out and crying at their desk come Monday morning.

How This Book Is Gonna Roll

In this book you will discover:

* The cultural and social barriers that keep us physically isolated.
* What touch does to the physical body and why it's so important.
* How much touch you prefer.
* How to navigate interpersonal boundaries with respect and consent.
* A framework that helps to keep your intentions clear.
* How to approach your friends in a thoughtful way.
* What items you will need for a successful encounter.
* Different ways to give and receive touch.
* What sort of touch you enjoy.
* Ways to share nurturing, human touch across a variety of situations.

✳ How to get your touch needs met by strangers in a safe, sane way.

✳ How to say no when you don't want to be touched.

✳ How to deal with boundary crossers.

PART I focuses on the role of nurturing human touch in our bodies and culture. PART II is a step-by-step guide that will teach you how to share touch with people you already know. Part III talks about moving through the world as a touch-aware person, and how to integrate touch into a variety of life situations.

You will also find a few activities sprinkled through the book (they say "Try Me!") that allow you to become more aware of the sense of touch, and play with some of the concepts we are discussing.

A Note About Pronouns

Did you know that "they" as a singular pronoun was voted the 2015 Word of the Year by the American Dialect Society? While its use can be traced back to Shakespeare, it has become popular among those who don't identify with binary gender roles. It is especially relevant for this book, and our examination of nurturing human touch, because for many people, being touched by someone of the opposite sex equals a romantic or sexual relationship.

By using "they" throughout this book, we can further contemplate the radical idea that touch doesn't have to be about sex, and thus doesn't need to be shared with someone of a specific gender. "They" as a singular pronoun is inclusive, neutral, and forward-thinking. All the cool kids are using it these days, and while I was never a cool kid, I will never stop trying to get into their clubhouse. "They" it is!

CHAPTER 2

What Is Touch Hunger?

Touch Hunger, or skin hunger, refers to an emotional response to lack of human touch after a long period without physical contact from another human. The term was coined in 1999 by Phyllis Davis in her book *The Power of Touch*. It is an apt description of a condition that affects young and old, rich and poor, black and white, liberal and conservative, and religious and atheist alike. You can have millions of dollars in your bank account, or thousands of friends on Facebook, and still suffer from touch hunger.

Touch hunger has many cultural causes – we'll be taking a deep dive into them in the next chapter. Some people hear the phrase and instantly know that is the thing that has been bothering them. Others might be surprised to learn that they have it, but when they start to think about it, it makes sense. Many others are aware that they don't get enough touch, but have rationalized away their need for it by trying to shut it down with mental logic.

Touch Hunger Is a Misnomer

While the term "touch hunger" refers to physical, skin-on-skin contact, it goes much deeper than that. Skin on its own is

> "One thing I notice is that I feel pretty invisible a lot of the time. I also tend to get into a mindset where I see myself as living in a sort of state of siege. Like I am a warrior braced for battle. This, in turn, may color the way I interact with others, which causes a circle further alienating myself.
>
> As for what I miss most about touch, I miss feeling connected and seen. I miss feeling cherished just for being, instead of what I can give, provide, or do for others. I miss feeling worthy of affection, rather than like I am begging for it."
>
> — MM, Single Parent

meaningless. Underneath the physical contact is a yearning for connection with other people. It's the desire to know that you are a valued member of the community, and that your presence matters. It's the need to be seen and heard, to be comforted in times of sorrow, to know that when things are rough, you are supported and not alone. When someone is touching you tenderly, you belong.

How Do You Know If You're Suffering from Touch Hunger?

Touch hunger has some surprising symptoms. It is a mystery why science has not focused more on the role of touch in overall health and wellness. Touch is integrative and affects many different systems in the body.

If you are suffering from touch hunger, you may have one or more of the following symptoms:

* You don't sleep well.
* You are distrustful of people.
* You have little to no physical contact with others.
* You are jumpy and easily startled.
* You are depressed.
* You are anxious.
* You are lonely.
* You feel disconnected.
* You can't remember the last time you had a good, long hug.
* Your body is tense.

While many people have underlying biological and neuro-logical reasons for disease and illnesses, touch hunger does exacerbate them. And while nurturing human touch will not cure you, it is a simple way to assuage physical and emotional pain.

8 Hugs a Day?

There's a popular Internet meme about hugging that makes the rounds regularly:

"We need 4 hugs a day for survival. We need 8 hugs a day for maintenance. We need 12 hugs a day for growth."
— Virginia Satir

Satir was a therapist who did pioneering work with families, and this quote has inspired thousands of families to be more physically affectionate with each other. Hugs are a wonderful, simple form of healing, but many of us already get hugged on

" I regularly have to fight the feeling that I'm no longer relevant or worthy of a simple hug. Depression is more inwardly expressed and ultimately more poisonous. Those moments when I could really use a hug or a pat on the back, my inward thoughts really do turn to feeling older, thinking in pointless circles.

Sometimes I wonder what others actually think of me. But without actually being held I have to wonder if words are just lip service. It all fuels the paranoid negative thinking that goes with depression and general anxiety disorder. It even adds to my physical pain when my muscles become rigid and I grind my teeth. It literally begins to feel like I'm in a monastery. It's harder to express myself in any fashion, not just physically."

— SR, Disabled

a regular basis. We still feel touch-starved, especially if we are single. We crave touch that is more nourishing, focused, and tender. Many of us suffer from a lack of opportunity be cared for, and to experience deep relaxation amongst people who see us as their family.

The Allegory of the Long Spoons

The allegory of the long spoons is a famous parable told in many religions. In the afterlife, people are sitting at a table with a lavish banquet, but all the utensils have long handles. In hell, people are starving because they can't get the food into

---

TRY ME! At your fingertips

Would you like to get acquainted with your own sense of touch?

Walk around your home, or office, and touch as many different objects/surfaces as you can with your eyes closed. Close your eyes AFTER you walk up to the object(s), not before. You can do this with your closet, your bathroom, and your refrigerator.

Check out the different textures of different objects. Can you sense if the object is warm or cold? If it's made from synthetic or natural materials? Are there surfaces that are pleasant or unpleasant to touch? Does this exercise help you to get a better feel (no pun intended) for how you might touch other people?

There's no need to do anything with this information; you don't need to write out your answers. Just consider it....

---

their mouths. In heaven, people are satiated – because they have learned to feed each other.

Touch hunger is much the same way: We are not meant to feed ourselves. If we are lucky, we might meet a person who will sit across the table from us and shovel food into our mouths between their other tasks. For a short period of time, we might have children who feed us. But while many of us are surrounded by people who might be willing to feed us, we are too afraid to ask, and thus, we starve.

I created this book to teach people how to use their long spoons to feed each other. There is no answer to touch hunger other than giving and receiving nurturing human touch, but I will let you decide for yourself. If nurturing human touch makes you feel happier, healthier, more relaxed, and grounded, then there's no reason not to do it. (Unless it's willfully non-consensual. Then it's not nurturing. This is an important distinction that we will discuss more in Chapter 6.)

Several years ago I realized that relying on my long spoons alone would not be enough to feed all the hungry people in the world. We could do Karuna Sessions 24/7, and there would still be a huge line outside our door. It was imperative that I teach others how to feed each other. And goddamnit, that's what I'm gonna do.

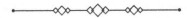

You've just learned about the concept of touch hunger: what it is, and how it can affect you physically, emotionally, and mentally. Touch can make you feel valued and connected to other humans, and lack of touch can make many health problems worse. Next, I'd like to discuss the cultural factors that serve to keep us isolated from one another.

Top 10 Cultural Myths That Discourage Nurturing Human Touch

You may have bought into some of these myths in the past, or been in situations where you were constrained by them. When you are able to see them for what they are – ideas that keep people separated – you can begin to make different choices based on what you want and need. Let's look at some of the biggies.

Number 10:
You're weak if you need others.

John Donne famously said that no man is an island, but American culture has been built around the myth of rugged individualism. You are supposed to be able to provide everything for yourself and your family, and if you are sick or poor, it's an indication of some great moral failing on your part, and you "deserve" to suffer.

"Pull yourself up by your bootstraps" is a common refrain, but it has become harder for many to obtain a decent pair of boots in the first place. Asking for help requires us to swallow our pride, or admit that we are a failure. Many of us prefer to despair alone instead of confessing vulnerability.

Number 9:
People touch each other too much already.

I've seen so many iterations of this sentiment written by people who don't like to be touched for a variety of reasons. They can't stand it when people hug them, and they condemn the entire culture, demanding that the world conform to their needs as opposed to clearly stating their own boundaries.

Your preference for touch will be quite individual, and might change over the course of your lifetime due to desire, health, and circumstances. The key is figuring out whether you need more or less touch, and then asking for it and/or saying no to it.

Number 8:
You will go to hell if you indulge in pleasure.

You should avoid touch because it feels good, and you should definitely be ashamed of yourself for enjoying it. Religion has waged war on bodily pleasure ever since Eve partook of the tasty fruit of the tree of knowledge. From hair shirts to draconian abortion laws, the message is loud and clear: Feeling good in your body is wrong, and you're going to hell if you pursue the sensual.

Gratification of our desires can only be had by worshipping an invisible being in the sky, and he hates competition. While our society becomes more secular by the day, this long-held belief that pleasure is bad continues to fuck with our heads, even as sex is used to sell everything. Guilt and shame have ruined countless sex lives and cast a dour pall over our culture.

Number 7:
Touch is for children.

When you are young, you might not think twice about touching your family members and friends. Your opportunities to receive nurturing touch vanish the older you get. With the average age of marriage rising, and more of us living alone, we are more physically isolated than ever.

Nurturing human touch is for children, but it is also for teenagers, twenty-somethings, people in their 30s, members of the AARP, and the elderly. If you don't believe me, spend an afternoon at a retirement home holding hands with the residents. You will have people lined up to receive a little human contact.

Number 6:

Your problem is mental, not physical.

Therapy has been a lifesaver for many people. It's great for teasing out the factors that make you unhappy and unfulfilled, but it's impossible to solve all of your problems from the neck up. Add in the stigma against seeking out assistance with our mental health, our lack of comprehensive health insurance and the high monetary cost associated with it, and, well, therapy really isn't going to fix everything.

Because therapists are ethically barred from touching their clients, there is a need for complementary therapy that involves physical touch. Many licensed therapists recognize this limitation as well, and are now beginning to send their clients to touch professionals.

Number 5:

Pets will fulfill your touch needs.

Animals, while wonderfully supportive and entertaining, will not fulfill your touch hunger. My cat can't hug me back, and if I asked for a specific sort of touch she would look at me blankly.

Not everyone can have a pet. You might live in a place that does not allow pets on the lease, or you can't afford kibble and vet care. You might have allergies, or you travel a lot for work and do not have the time to care for an animal. Or maybe you do have a pet, but still miss nurturing human touch. Pets are not a substitute for other humans.

Number 4:

You can only touch your romantic partners.

It's a wonderful thing to have a lover who likes to hug and snuggle, but what do you do when you are single? There are many of us who are lacking opportunities to get our touch needs met because we don't have (or, in some cases, want) a romantic partner. This paradigm is leaving millions of us without nurturing human touch.

What if you are one of the millions of people who are deemed undesirable because of your appearance, age, disabilities, or physical, mental, or emotional health? What if you're painfully shy, or you're working two jobs and don't have time to date? And what do you do when a relationship ends, through death or divorce, and you're not ready to date again? How do you get your touch needs met when you are grieving (arguably one life situation where nurturing human touch is most beneficial)?

Number 3:
You'll get cooties if you touch other people.

If you live in a first-world country, you probably shower daily and engage in various forms of hair removal and odor abatement. Nevertheless, your body is still laden with internal and external bacteria. Sometimes your body contracts viruses or infections that can kill. In addition, your body smells funny to others. You fart, burp, sneeze, and cough, and your body excretes sweat, mucus, urine, and feces. The yuck factor is strong with humans.

The health benefits for touch are just as compelling as the health benefits of avoiding others. If your immune system is compromised, then by all means be cautious, but you can decide for yourself whether it's worth the risk.

Number 2:
Sex is the only way to get your touch needs met.

Touch is a component of sex, but sex does not have to be a component of nurturing human touch. You can have a wonderful time touching people platonically without involving anyone's genitals. Many of my single women friends would ask my ex-boyfriend to cuddle because they felt safe with him – they knew he wasn't going to try to turn a snuggle into sex (or recruit them for a threesome). The key is creating clear boundaries and intentions from the get-go.

Because our culture has deemed sex as the only delivery device for tender touch, millions of single people engage in hookups, even though it's not exactly what they are looking for. Sometimes it can be fun and satisfying, but other times it can leave you feeling lonely, sad, and unfulfilled.

Number 1:
Your partner will fulfill all of your touch needs.

Nurturing human touch is something your partner may or may not have the bandwidth to give. The soulmate myth tells you that you will have one person who will be your everything: your best friend, your co-parent, your sex partner, your confidante, your activity partner, your cheerleader, and your supporter. That's a lot of responsibility to squeeze in after going to work, taking care of your family, pursuing your passions, and keeping your body and home in good shape. And of course there is nothing lonelier than being in a sexless marriage – couples who aren't having sex usually aren't sharing nurturing human touch.

We've done many a Karuna Session with married people whose partners sleep in the furthest corner of their shared king-sized bed and don't like to snuggle after sex. We've also done sessions with married folks whose partners are touchy-feely, yet they still find it beneficial to receive touch in a situation where their needs are most important, and they won't be expected to reciprocate with sex.

## CHAPTER 3

The Culture Surrounding Touch

"**Y**ES!" YOU SAY, "This whole touch thing sounds really cool! I'm all in!"

Slow down there, little buckaroo! Before you can create new touch relationships, you will need to run through the mental obstacle course that our culture has created. Grab some popcorn, and settle in: There are a lot of factors that keep us separated.

A famous study done in the 1960s by Sidney Jourard[1] observed friends in cafes around the world as they physically interacted with each other. Puerto Ricans had the highest amount of touch, with 180 touches in an hour. The French touched each other 110 times.

Americans? They touched each other twice. Once, twice. Two times. This research shows that touch is dictated by culture and varies wildly around the world.

In this chapter, I'd like to examine some of the conditions that breed touch hunger in the United States. I know that many of my readers are forward-thinking and like to rebel against the mainstream. Knowledge is power, and shifting your perspective will allow you to figure out what you want instead of following someone else's rules.

Touch in the Past

Pre-history

When we were living in caves, as hunter/gatherers, we slept in big, tangled piles of bodies. There was no separation of any sort: Men would lie on other men, or next to their full-grown children. We did it for two primary reasons: warmth and safety. And, if studies of primates[2] are correct, this also fostered co-operation and sharing, an important thing when resources are scarce.

To be ostracized from your tribe, to be alone, isolated, and solitary, meant death. Humans needed other humans in order to survive. We still do.

Agrarian life

When we switched from primarily being hunter/gatherers to farmers, we had a lot of physical contact with our family members. Before the advent of electricity, people would sleep in one room, often in the same bed. Staying warm was a matter of survival, and when fuel for fires was expensive and scarce, bodies were a great source of free heat. Amongst very poor families, livestock would also sleep indoors to provide more warmth.

Hygiene in many places was a luxury (no electricity or indoor plumbing = heating hot water on the stove for bathing), though many cultures had bathhouses where people would bathe together.

In addition, doctors and hairdressers were few and far between – we cut each other's hair and changed bandages and bedpans when our family members were sick and injured. Animals provided food, not companionship. Death was much more sudden, and common, and the dead would lie in our parlors, and be washed and prepared for burial, by people who loved them

Romantic relationships

The definition of marriage has changed many times over the years, but for much of civilization, it was a business contract between two families. Prior to marriage, men and women were isolated from each other, socializing primarily with their own gender. Young couples who married often lived with their husbands' families. Sexual compatibility was a crapshoot, and privacy post-wedding night was a rarity. Because parents often shared a bed with their children, sex was less private.

Childbearing and -rearing

Until quite recently, children were frequent, and unplanned. The age of marriage was well south of 20, and it was not uncommon for women to have 10 children. It was also not uncommon for children to die during childbirth, or before the age of 10. Those who survived often started working on the farm long before they reached adulthood, or worked outside the home to support the family.

While some families raised their children with a great deal of touch and warmth and closeness, many others mentally, physically, and sexually abused their children. In more recent history, child-rearing experts advised parents not to hold their babies when they cried, and to have them sleep in separate beds.

Touch in the Present

Things have changed immensely in the past 100 years, and nowhere is this more evident than the United States. Let's look at some common aspects of 21st-century life and how they have curtailed our opportunities for touch.

Stratification and separation

Whereas in the past we would live in multi-generational

households, in small spaces, 27% of us now live alone.[3] We have preschools and retirement homes where our youngest and oldest spend time only with people their own age. The average number of children has dropped from 3.8 in 1957 to 1.9 in 2010,[4] and due to our utter lack of support for child-rearing in the U.S., infants as young as six weeks are shuttled off to child-care facilities while their parents work to support them. Almost every family requires two incomes to survive.

Many of us move thousands of miles away from our families after college, leaving our children without the love, support, and guidance of their grandparents. Grandchildren and grandparents benefit from the tender touch shared in these relationships, especially the elderly whose spouses have died.

Even when we have relatives close by, the average size of our homes has increased to 2,392 square feet in 2010 (up from 1,660 in 1973),[5] giving us more opportunities to be isolated from our own families.

Loneliness

All this isolation is not good for our physical bodies. The World Health Organization has declared that loneliness will be a public health epidemic in the coming years. It contributes to inflammation, heart disease, and cognitive decline, and can be more damaging than a lifetime of smoking.[6] This trend will undoubtedly continue, leading to higher costs in medical care and greater strain on families.

Americans spend an average of 18 hours a week watching TV[7] and two hours and 37 minutes a day on their smartphones.[8] While we can now keep up with our classmates from high school and college on Facebook, getting together with friends in real life can be next to impossible. In 2011, we reported an average of 2.08 friends we could call in a crisis, down from 2.94 in 1985.[9] When we're sick and need food and medicine, we call Amazon, not Allan.

Busyness

Most of us are scheduled to the gills, and maintaining relationships is a mammoth task. Our day-to-day lives are busier than ever, filled with endless to-do lists, goals, and priorities. While the advent of electricity and home appliances like dishwashers and washer-dryers promised us more leisure time, we've quickly found ways to fill that time.

Stuff

While tiny homes have become trendy, many Americans live in large suburban homes filled with possessions, gadgets, and clothes. Retail therapy is a thing: Women in particular try to soothe themselves by buying a new pair of shoes or a lipstick. Almost 1 in 10 Americans rents storage space to hold their extra junk, according to the Self Storage Association. And while acquiring and managing all of our things takes up a lot of time, money, and attention, we're still not happy.

Information overload

Scientists estimate that our brains receive 34 gigabytes of information a day(!).[10] Getting to an inbox with zero emails has become a holy grail of sorts, and everywhere we look, crises, scandals, and issues demand our attention. Deep thinking and unplugging from the grid has become increasingly difficult. All this information keeps us distracted, and many of us reflexively withdraw into our own narcissistic bubbles to protect ourselves from overstimulation.

Research has also shown that rates of depression and suicide have risen with the ubiquity of smartphones.[11] Teenagers are having fewer real-time social interactions. Even that long-time favorite teenage activity – sex – has dropped by 40% since 1991 for kids in the ninth grade(!).[12]

37

Mental health

For many of us, the inside of our head is a bad neighborhood these days. In 2016, 16.2 million American adults experienced a major depressive episode, and 9.8 million of us had serious thoughts of committing suicide.[13] Anxiety affects nearly one in five people.[14] Many of us are in pain: We spend $78.5 billion annually on lost productivity, healthcare, rehab, and criminal justice because of prescription opioid use.[15] Despite being a wealthy, first-world country, the United States is 108th out of 140 on the Happy Planet Index.

Pets

The popularity of dogs and cats as house pets has risen dramatically in the past 50 years. In 2016, we spent over $66 billion[16] on caring and feeding for our house pets. People often refer to their animals as their "fur babies," dress them in human clothes, and sleep in the same bed as them. Oftentimes they are our sole source of intimacy and physical contact. (In 2015 the media was abuzz with reports of how staring into your dog's eyes can give you an oxytocin boost.[17]) While they have been a godsend for lonely adults, they are no substitute for other humans.

Romance, sex, and love

In the United States, 45% of us check the "single" box on the census form,[18] and marriage is no longer a given. In 1960, 59% of adults aged 18 to 29 were married; in 2010, that number was 20%.[19] With the advent of birth control and women working to support themselves, relationships have experienced a dramatic shift. While arranged marriages still exist, most people now look for their soulmate, The One. Despite many Christian churches and sex ed curriculums demanding abstinence, many of us engage in hookup culture. And with the advent

TRY ME! A Spy in the House of Touch

Go to a public area: a park, a mall, a subway station, a bar, a grocery store, a coffee shop. Find a place to sit or stand unobtrusively, and observe people and their physical interactions (or lack thereof). Spend about 10 to 15 minutes watching.

Can you guess the relationships between people depending on how often they touch each other? Is it obvious when people are trying to avoid touch? Is there a difference in the frequency/quality of touch depending on people's gender or age? Does the amount of touch vary depending on people's purpose for being in that space?

Extra bonus points (which entitle you to absolutely nothing) for going to a place where you can observe another culture (an ethnic grocery store, or a mosque, for instance) and see how their culture around touch differs from American culture around touch.

of online dating, we operate under the delusion that there is a limitless pool of possibilities, and have become much pickier, rejecting potential mates for something as simple as liking the wrong sports team.

Masculinity

Men struggle with touch hunger more than women. After puberty, the most acceptable way for men to get touch is through their sexual relationships. At the same time, by casting men as sexual predators and women as prey, we have pathologized casual male touch. The result has been that unpartnered men are more touch-deprived than women, and have fewer ways to get their touch needs met.

As children, boys touch each other easily and frequently, but as they age, sports and war are two of the only arenas where it is acceptable to be physical with the guys. Many men begin to awaken to the joys of nurturing human touch when they become fathers, but touch hunger is rampant with those who are single. The "bro hug" with a pat on the back is as far as most men get with other men. And while many women are touchy-feely with their women friends, most men don't have the same level of physical ease with other men.

When there was no sex before marriage, and society was more gender segregated, men were more affectionate with their male friends. Photos from the early 1900s showed men photographed with their friends, co-workers, and teammates, embracing each other tenderly and without shame. Men who were soldiers would frequently share beds.[20] While homosexuality existed, it was considered an activity, not an identity, and touching another man didn't equate to same-sex attraction.

In many parts of the world today, men share physical affection, and can often be seen walking down the street holding hands. Many American men scoff at the idea of sharing nurturing human touch with their male friends for fear of appearing weak, effeminate, or gay. Sharing nurturing human touch with other men isn't the only option, but it is an option.

As you can see, Americans have become more isolated over the past 5/50/100 years. We have adopted lifestyle patterns that keep us from being in contact with other people, and our health and sanity are suffering. In the next chapter, we'll explore why touch is so good for your health.

Top 10 Reasons to Engage in Nurturing Human Touch

Why would you deliberately and consciously seek out more touch in your life? Here are 10 compelling reasons to do so.

Number 10:

You get off your phone.

One of the most positive things we do during a Karuna Session is get you off your phone for 90 minutes. Most of us spend a lot of time in our heads, and not much in our bodies, thanks in large part to mobile technology. Touching other people instead of your smartphone helps you slow down and take a break from thinking. You can connect with people you love...and it doesn't involve texting them incessantly.

Number 9:

You will sleep better.

Sleep is an integral part of any health regimen, and getting good, nourishing sleep is much better than being up at 4 a.m. thinking about all the problems of the world. Thanks to the sleep monitor in my FitBit, I have discovered I get better sleep on the days I receive a lot of touch. You will too.

Number 8:

You get to play.

Getting together with your friends for nurturing human touch will give you an opportunity to hop a time machine back to a time when things were less complicated. Almost everyone

I know is overscheduled. We're happier and less stressed when we make time to play, and play enhances cooperation and bonding. It's a great way to take a break from all the heaviness of today's world (also known as the "default world") and create some space to enjoy being human.

Number 7:
You are comforted.

Most people feel scared and powerless about personal and collective struggles at many times in their lives. It's a glorious feeling to have another person hold you for just a moment as a way of saying, "It's okay. You're safe. I'm here, and I've got you." We can all use a bit more of that right now...and you don't even have to admit you feel scared.

Number 6:
You will feel more attractive.

When you haven't been touched with tenderness for a long time, your desire for it can be overwhelming. Nurturing human touch will help you relax and feel more comfortable while dating – you can be less rushed and pressured when looking for sex or romance. You can take your time getting to know people. You can engage in hooking up because you want to have sex, not because you're suffering from a lack of touch.

Number 5:
You can give/receive support.

Our culture has a poor relationship with painful emotions. Most of us don't know how to be empathetic when someone is suffering. A long, heartfelt hug is worth 100 sympathy cards. It lets people know they are seen, loved, and valued when they aren't at their best. Plus, you don't have to struggle for just the right words, and risk saying something hurtful.

Number 4:

You don't have to pay for it.

Nurturing human touch is the most economical of wellness practices. You can join or quit at any time! No monthly or annual fees required! You don't have to buy special food or supplements, hire a trainer, or sign up for a class. There is a price – you will have to become vulnerable, ask for what you want, and risk changing the dynamics of your friendships – but that is a paltry sum.

Number 3:

You don't need special training or equipment..

Most health and wellness practices take time to master, and many of them require an investment in special clothing or accoutrements. Not so for nurturing human touch; the learning curve is short, and gentle, and this book will provide you with the training you need. Touch is a gentle way to move your body if you're out of shape, you have mobility issues, or you're recovering from an illness.

Number 2:

Your health will improve and you will look younger.

Our bodies need touch and oxytocin, especially as we age. The next chapter discusses the physiological health benefits at length, but nurturing human touch can have a positive effect on several systems of the mind and body. People often look 10 years younger after we finish a session with them; their bodies soak up nurturing human touch like the desert soaks up the rain after a long drought.

Number 1:

You will feel good.

When nurturing human touch is consensual, it will make

you feel better. Feeling good is the goal of most health and wellness regimens. Touch gives you that in spades!

While it won't cure your cancer or obliterate your depression, it will give you some relief. And it's much more pleasurable than staring at a plate of broccoli. (That's not an excuse to pass on the healthy eating. Your mom told you to eat your broccoli, and she was right. Eat your broccoli.)

## CHAPTER 4

The Health Benefits of Touch

TOUCH IS AN inherent part of our biology. We evolved with touch as a species, and it plays an important role in our development as infants. It is a universal human experience. Our primate cousins can often be observed grooming each other, and spend 20% of their time in this activity. While scientists originally thought that the purpose of this was to keep themselves free of vermin, researchers now speculate that they also groom in order to foster cooperation and sharing.[1]

The Role of Touch in Human Development.

Touch is the first sense we develop, after eight weeks of gestation. In recent years, "kangaroo care" (the practice of parents holding their babies with skin-to-skin contact) has become popular. Proponents link it to weight gain, better sleep patterns,[2] and better bonding between infants and their caregivers. Infant massage has also been found to be beneficial for both infants and parents.[3]

Harry Harlow's study of rhesus monkeys, who clung to a cloth "mother" instead of the wire one that gave food, shows just how strong the need for touch is.[4] Indeed, the desire for comfort and connection is as compelling as the need for food. Nurturing human touch belongs on every level of Maslow's Hierarchy of Needs: physiological, safety, love/belonging, esteem, and self-actualization.

Infant cuddling for preemies has been found to help them gain weight, allow their immune system and organs to develop more quickly, and get them home to their families earlier.[5] It is a commonly-used tool in neonatal intensive care units around the country with prematurely born and sick infants. There are legions of volunteers whose sole job is to cuddle and massage those humans who make an early entrance onto the world stage. Infant cuddling is also being used extensively with babies who are born with drug addictions.[6]

During the first two years of life, our brains continue to develop neural pathways. Being frequently held and touched allows the parts of the brain that deal with emotional regulation and safety to become robust, and provides us with a strong foundation for managing stress and relationships as we age.[7] Attachment parenting gives little humans the security and confidence they need to explore the world and take risks.[8] In a follow-up study more than 10 years later, consistent touch given to infants was found to have long-term developmental benefits.[9]

Touch is also a profound and universal method of communication. In an experiment by the Greater Good Science Center at UC Berkeley, it was found that a one-second touch can convey a range of emotions to the receiver.[10] The study found that receivers were accurately able to identify a range of emotions (e.g., sad, surprised, or happy) from a touch given by a stranger without any other cues. Actions really do speak louder than words.

The human body is equipped with a somatosensory system that registers changes on the surface of the body. Our skin is our largest organ, covering 22 square feet if laid flat and weighing 20 pounds for an average size human. There are four different types of cells on the skin surface that respond to different types of touch. The Merkel cells respond specifically to human touch,[11] and our hands are equipped with tactile corpuscles that make it feel good to touch.[12] It's a win-win, for the giver and the receiver.

Oxytocin and Physical Health

Touch also releases a hormone known as oxytocin. It's made both in the brain and the bloodstream. It was originally associated with childbirth – the word "oxytocin" is Greek for "swift birth." Synthetic oxytocin, or pitocin, is given to laboring mothers to facilitate the birth process by making the contractions come faster. Oxytocin also helps the mother and her offspring bond during nursing.

In the past several years, oxytocin has been extolled for its psychological effects as the "moral molecule" or the "love hormone," but its effects on the physical body are well worth examining. Among other things, oxytocin has been found to:

* Lower heart rate and blood pressure[13]
* Counteract the effects of cortisol in the heart[14]
* Help muscles regenerate[15]
* Boost the immune system[16]
* Accelerate wound healing[17]
* Reduce inflammation[18]
* Aid in pain relief[19]
* Counteract the effects of alcohol[20]
* Reduce appetite[21]

That's a damn fine list of effects that correspond with health, wellness, and vitality.

TRY ME! Self-Soothing

Remove all rings, bracelets, watches, etc. With your right arm by your side but slightly forward, turn it so the palm is facing up, and it is bent at a 75- to 90-degree angle. Take your left hand, and put it above your right hand, with the fingertips touching. Begin moving your fingertips on your left hand down the fingers, across the palm, over the wrist, and up the forearm of your right hand and arm. Stop just before the elbow. Return the fingertips of the left hand to their original position, touching the fingertips of the right hand. Repeat five times.

When you've finished, answer these questions. How did it make you feel to touch yourself tenderly? Was it relaxing? Comforting? Did it feel foreign? Were you angry that you had to touch yourself instead of having someone else do it? Did one hand or the other feel like it belonged to someone else and wasn't part of you at all?

Scientists are still learning about oxytocin, but most of the experiments done are in labs with nasally administered oxytocin spray. To me, this is counter-intuitive and wrong. I get that it's easier to run experiments in a lab with a carefully measured, medicalized substance, but oxytocin should not be divorced from humans and human touch. When you can generate your own oxytocin through nurturing human touch, you have an abundant – and free! – source.

The Role of the Vagus Nerve

The vagus nerve runs throughout the torso, from the brain to the genital area. It touches most of our major organs. Often labeled the "compassion nerve," the vagus nerve is integral to the body's ability to relax and feel safe. Research has found that it can affect both inflammation and the immune system, and can positively influence metabolism.[22] It can be stimulated via touch to the back, neck, and shoulders.

The vagus nerve is part of the parasympathetic nervous system. The parasympathetic nervous system regulates the "rest and digest" cycle. It's what allows us to calm down after a dangerous experience, to take us back from Defcon 5 to no present danger. This cycle can be quite useful if you're protecting yourself from hungry wild animals, but with the stresses of modern living, we often find ourselves remaining alert and vigilant, always on guard. We find ourselves with fewer opportunities to get into a relaxed state, and it takes a toll on the physical body.

Physical proximity and contact of our tribe members would have been the best way to tell ourselves that we were safe from danger, and not alone. Is this perhaps what we have been missing?

Be Your Own Lab Rat

The research on oxytocin and touch is ongoing, but the current knowledge is compelling. There is robust literature on the benefits of human touch. More importantly, it is integrative and comprehensive: It is not only good for the physical body, but for emotional and psychological health as well. The feeling of safety and calm that can be created by the touch of trusted humans is profound. Touch may well be the next thing doctors prescribe for anxiety, addiction, depression, or PTSD, but the beauty part is that you do not need to ask your doctor for a prescription to see if it will work for you. You will be surprised at how much better you feel!

You just learned about how touch positively impacts the human body, and you want to get more of it into your life. But is it possible to have too much of a good thing? How do you know how much touch you need? We'll be covering that in the next chapter.

## CHAPTER 5

How Much Touch?

ABOUT HALFWAY THROUGH my nine-day stay at Burning Man in 2017, I tried to figure out how many people I had hugged. Dozens? Hundreds? It was hard to tell. Our camp was in a high-traffic spot and had a bar, so I was meeting lots of new people every day.

If we live in a touch-hungry culture, Burning Man is an all-you-can-eat buffet where people pig out. One oft-cited reason for attendance is to connect with like-minded people, and participants tend to be more friendly and open than they are in their day-to-day lives. A hug is a standard greeting for old friends, friends-of-friends, and strangers. I had a wonderful time sharing my hugging superpower with people, and I had more than one grateful person tell me that I gave amazing hugs.

My campmate, Morgan, was not amused. They are a touch-sensitive person even with people they know, and all the hugging started to get to them. By the end of the week, I saw them sticking out their hand for a handshake when a stranger introduced themselves. While I found the level of physical interaction to be comforting and relaxing (I did notice a lack of consent, but that's a whole different topic), Morgan became more agitated and guarded.

Which one of us is right? Is Burning Man a paradise where

you can get your touch needs met? Or a torture chamber where the punishment comes in the form of hugs? Both of our responses are correct, real, true and equally valid. Touch is vital for our health and wellness, but everyone has different needs and desires for physical contact with others. It's important to figure out how much touch you need before you can involve other people to get said touch needs met.

You Say Tomato, I Say Tomahto

Touch is not a one-size-fits-all need with a single solution. If we frame it as a component of wellness, we can approach it the same way we approach other aspects of wellness.

Exhibit A: Diet. Scientists and nutritionists can't decide what constitutes a healthy diet, and even if they can, their guidelines don't work for everyone. Some people thrive as vegetarians; others avoid gluten. Some people are content to eat frozen dinners popped in the microwave; others prepare elaborate meals from organic ingredients three times a day. What works for you might not work for your children or your friends, and your own needs will likely change over your lifetime.

Similarly, we all have varying needs for nurturing human touch. The most important thing is to figure out how much touch is right for you, and to be comfortable with both asking for what you want, and creating strong boundaries around what you don't want.

If you're reading this book, it's either because (a) you already know you want more touch in your life, (b) you're curious about how nurturing human touch can improve your life, (c) you'll try anything to make yourself feel better and healthier, or (d) all of the above. Regardless of your reason, you will need to get acquainted with your own desires.

The clients I've encountered through Karuna Sessions came

to see us because they knew they wanted/needed touch, or a therapist/friend told them they would benefit from our services. Still, they had trepidation around it. Once they worked through all the mental weirdness of being touched by strangers, and allowed themselves to receive touch, they felt happier, calmer, grounded, and more relaxed. They also often walked out realizing that touch is something they need, and tried to figure out how to get those needs met on a regular basis without our assistance.

Think about experimenting with nurturing human touch as you would contemplate trying an exotic food: It seems a bit weird, but you're going to give it a try because you're open-minded. It may be love at first bite, or you may require several bites. Once you try it, you're welcome to say "No, not for me." But I do encourage you to taste it before rejecting it outright.

How Do I Know What My Touch Type Is?

Just like some people have high sex drives and others have low sex drives, people have different needs and desires around touch. But how can you determine what your own needs are? I decided to go all Cosmopolitan on your ass and put together this handy-dandy quiz to help you figure out where you fall on the spectrum. Please note that this is highly unscientific and meant only to give you a starting point for reflecting on how you – and others – interact via touch.

The How Much Touch? Quiz

1. Someone walks up to behind you and puts their hand on your shoulder. You:
   A. Don't notice their hand until they tap you on the shoulder a few times, then you turn around and give them a hug and greet them warmly.

B. Move away from their hand, and figure out whether you know them before deciding how to proceed.

C. Flinch and shrink back from their touch.

D. Instinctively move away when their hand is six inches from your body because you felt them coming.

2. You go to a party where several of your friends are present. You:

A. Run up to everyone, even distant acquaintances, and give them a big hug.

B. Look for the people you know and feel safe around, and give them a hug.

C. Hold out your hand for a handshake, as people approach you.

D. Stand in the corner with your hands full of food and drink, chat with the person you came with, and pray that no one wants to hug you.

3. You and your partner split up. When you find yourself faced with an empty bed every night, you:

A. Hop on Tinder and look for hookups and try to get them to spend the night after having sex.

B. Pull your cat under the covers and hug him while you sleep.

C. Get yourself a body pillow and wrap yourself around it.

D. Breathe a huge sigh of relief – you hated sharing a bed with another person.

4. You are at your company Christmas party. One of your coworkers has had a few too many Moscow Mules and starts hugging you and telling you how much they love you. Do you:

A. Hug them back and say "I love you too"?

B. Pull away as quickly as possible and give them a pat on

the back?

C. Dodge their hug and consider calling HR the next morning?

D. Freeze up and shut down?

5. You are attending a weekend-long yoga retreat. When you arrive, you:

A. Go up to everyone, introduce yourself, and give them a hug.

B. Chat with another attendee, and if you feel safe with them, hug them at the end of the first meeting.

C. Hang out with the person you came with and avoid the others.

D. Roll out your yoga mat, close your eyes, and start meditating.

6. You meet someone cute in your history class, and you start studying together. One night you start talking and discover that you have a lot in common. At the end of the evening, you want to give them a hug. You:

A. Grab them and give them a hug!

B. Ask them if they would like a hug, and follow their lead.

C. Decide to wait until you've had a couple more meetings to make sure they are interested in you.

D. OMGOMGOMG they're so cute and I want to touch them but I would probably freak out if I did.

7. A buddy wants to introduce you to a friend of theirs. You exchange a few text messages, and the three of you arrange a meeting. When you arrive, the new friend goes in for a hug from you. You:

A. Don't think twice about hugging them back!

B. Hug them a little bit, but pull away quickly.

C. Hold out your hand for a handshake instead.

D. Move out of the way and use your buddy as a human shield.

8. **Which of the following scenarios best describes your family:**

A. My large extended family was full of frequent huggers.

B. My parents, brothers, and sisters shared touch, and my parents let me choose whether or not to hug relatives.

C. My parents were busy, and physical affection wasn't shared.

D. Most of the touch was in the form of punches and slaps.

9. **You scored an amazing velvet shirt the last time you went shopping and decide to wear it out to a show. People keep coming up to you and petting your shirt. You:**

A. Are thrilled – you were hoping that people would want to touch the shirt! That's why you bought it!

B. Look them in the eyes and say "Didn't your mother teach you to ask before you touch?" and move your arm away.

C. Pull your arm out of their reach and walk off.

D. Find a place to stand where you can see people coming at you, and vow to get rid of the shirt the next day.

10. **An old friend comes through town with their four-year-old. When you meet up for lunch, your friend hugs you, introduces the kid, and tells them to give you a hug. It's clear to you that the kid doesn't want to hug you. Do you:**

A. Hug the kid anyway – they are so darn cute!

B. Crouch down so you are more on eye level with them and ask them if they would like a hug.

C. Hold out your hand and ask for a high five instead.

D. Give the child a knowing smile – you remember how horrible it was when your parents made you hug strangers.

Scores:

For every A, give yourself 4 points
For every B, give yourself 3 points
For every C, give yourself 2 points
For every D, give yourself 1 point

Results:

0-10: You are a touch-averse person. You are comfortable with being alone and may not understand social cues or body language. Being touched by another human being is overwhelming and uncomfortable. You will do anything in your power to avoid touch, and don't understand why people crave it.

11-20: You are a touch-sensitive person. You are highly aware of the breeze on your skin and the seams of your underwear. You may have been raised by people with poor boundaries. You don't like to be touched by 99.99% of people out there, and the .01% who can touch you only do so after being thoroughly vetted.

21-30: You are a touch-situational person. While you may crave physical contact, you are discerning about who you give to and receive from. You are highly aware of social cues and contexts, and will evaluate a situation to determine whether touch is appropriate.

31-40: You are a touch-enthusiastic person. You want to hug everyone all the time. You love touching and being touched and don't always pay attention to other people's boundaries. Touch calms you down and makes you feel good, so you naturally assume that everyone feels the same way you do!

All of these touch types are acceptable – I know people who fall into each of these categories. Regardless of where you fall on this spectrum, you will need to figure out how to navigate touch. **One of the biggest mistakes you can make is assuming that everyone else has the same boundaries and desires as you.** Navigating the space that separates humans from other humans is simple, but it isn't easy. You struggle with it because you aren't taught how to do it.

It is your responsibility to take care of yourself and set your boundaries with regard to people touching you. It is also your responsibility to ask for what you want and get consent before you touch other people. You will likely find yourself more frequently experiencing one side more than the other (boundaries vs. consent, or vice versa), but it's important to be well-versed and adept in handling it from both sides.

"Gee willikers, Miss Jordan, I have no idea of how to do that!" you are probably thinking to yourself right now.

Yep, I know. Believe me, I know. So let's arm you with the tools to do so.

CHAPTER 6

Consent

THIS CHAPTER IS the most important part of the book. You will not be able to create relationships around nurturing human touch without it. Please memorize it, rip the pages out, chew them up, and swallow them so that they become part of the very fiber of your being!

(I'm kidding! Please don't eat your computer – it will give you wicked awful indigestion.)

You will be going through a series of exercises later in the book that are designed to give you practice with consent. But for now, please spend some time contemplating these concepts before you embark on crossing the gulf that physically separates us from other people.

Choice

Nurturing human touch is pleasurable. It feels good to be hugged, touched, and caressed by another human being. *But unless it's consensual, it will not feel good. Period.*

My friend Riley is touch-sensitive. They recently talked to me about being at a pagan gathering where hugging was considered a standard greeting, and how much they hated it. None of the people who were coming up to them and hugging them noticed their body stiffening up and trying to pull away. Everyone wanted to hug them and didn't care whether or not they wanted a hug. That is not nurturing human touch. We want to make people feel better, not worse, and create a culture where we honor people's desires...and have them honor ours.

What Is a Boundary?

A boundary is a clearly stated limit. It is a line drawn in the proverbial sand that tells another person what is (un)acceptable to you. Boundaries define your personal space and territory, and create a container that keeps you safe when they are honored.

This may sound like a simple thing, but most of our personal relationships operate with assumptions, not boundaries. How many marriages have tanked because one person assumed that the word "marriage" meant the exact same thing to the other person? We frequently expect other people will know what it is that we want and need without us explicitly stating it. These assumptions lead to all sorts of problems.

The Fluidity of Boundaries

One of the most confusing things about boundaries is that they are constantly shifting. Sometimes the word "no" is a complete sentence. Other times it means "not right now," "not that thing," or "not with you." Boundaries can change very quickly, and what was okay earlier might no longer be okay a moment later. If you need more clarification about someone's boundaries, you should ask for it: Remember that while you are learning to hear yes and no, others are learning to ask for what they want and need. If you are negotiating boundaries with people with whom you have an ongoing relationship, your boundaries likely won't be static.

Rejection

Rejection is a fact of life, and yet, we're not taught to deal with it. This can have deleterious consequences: There are hundreds of stories about women being killed after they say no to a dude that's hitting on them. Glenn Close terrified Michael Douglas and moviegoing audiences in *Fatal Attraction*. Clearly, some people can't handle rejection.

We will be learning to hear and say the word "no" later in the book, but rest assured, as you try to get your touch needs met, you're going to be hearing this word. A lot. Get used to it, and always be willing to take "no" for an answer. My wise friend Francesca says this: A boundary is for me, not against you. Not only will you hear "no," you'll probably have lots of opportunities to say it. With practice it will get easier.

It's okay to ask. And it's okay to say "no."

Unpleasant Touch

We pay strangers to touch us in unpleasant ways all the time.

Every time I go to Dallas, I get a Korean salt scrub, a process in which almost every inch of my body is scrubbed with a rough mitt that removes about a pound of dead skin. While I love the way it looks and feels when it is done, it isn't much fun in the moment.

The medical industry is one of the largest and most expensive purveyors of unwanted touch. Gynecologists and proctologists and mammograms, oh my! Blood draws, surgical incisions, those long needles full of novocaine piercing your gums...you get the idea.

These are types of touch we consent to. That doesn't make them more comfortable.

Unwanted Touch

Standing on a crowded subway train, being jostled by smelly, sweaty people during evening rush hour. Squeezing in between people, trying to get close to the stage at a sold-out show so you can watch the bass player do their thing. Someone talking on their cell phone who walks right into you at the grocery store.

Most people never think about these types of unwanted touch. Some of them are merely the price of living in close proximity to other humans; other times they are deliberate or inspired by self-absorption. As you become more aware of touch and boundaries, you will begin to realize that many people are clueless when it comes to their personal space, and how they encroach on other people's personal space.

Non-Consensual Touch

Certain kinds of non-consensual touch can really fuck us up, and make us wary or sensitive to touch. One in six people are physically or sexually abused as a child,[1] and 75% of those experiences involve someone the child knew.[2] Many children will be physically bullied by their classmates for being different. One in five women will be raped during her lifetime, and one in 71 men will experience rape.[3]

Most people will never introduce themselves with "Hi! I'm Aaron and my uncle molested me as a child, nice to meet you!" These hidden traumas make it especially important to ask for consent before you touch another. Getting consent takes a moment, and will go a long way to make the receiver feel safe(r). Not getting consent can reinforce the idea that touch is bad, dangerous, and unpleasant.

We are experiencing a huge cultural shift around consent, and touch, right now. The media creates cultural narratives, but changing this particular narrative will happen through billions of micro-encounters between people every single day. You have the power to rewrite this story by asking for the touch you want and asking before you touch someone. You can also model good behavior for others that will give them permission to create better boundaries for themselves going forward, or cause them to pause before crossing someone's boundaries.

Uneven Power and Touch

We live in a hierarchical society where we have bosses, religious leaders, sports heroes, commanders, celebrities, teachers, and authorities of all stripes. Touch between people who are on different levels of power or influence can be uncomfortable, especially for the person who has less power. 2017 was the year of #metoo, when we finally began to talk about how ubiquitous this type of touch is, and how much damage it does in the workplace.

If at all possible, figuring out ways to get your touch needs met through someone in your peer group is the best solution. It's not necessary to avoid touch with people who have less power than you – there are many social situations where you will have such encounters – but be aware that while you may think it's fine to hug your employee, they may have a completely different thought process and experience around it.

Permission vs. Consent

This is an advanced concept, and one that people often confuse. It's also extremely important. It comes from Betty Martin, a Seattle-based sexuality and intimacy coach. It looks a little like this:

"Can I have a hug?" *vs.* "Would you like a hug?"

TRY ME! Permission vs. Consent

As you move through your week, practice getting consent before you touch someone, or ask them to touch you. Ask friends "Would you like a hug?" upon meeting up with them and see how they respond. You can also try asking permission – "Can I hug you?" – and see if they respond differently.

Getting consent to touch doesn't need to be specifically around nurturing human touch. Many women see a tag sticking out of someone's neckline and reach out to tuck it back in without saying, "Your tag is sticking out of your shirt – would you like me to tuck it in for you?" Asking for consent is a habit worth cultivating.

You can also ask people to touch you. "Would you please scratch my back?" "Would you touch my forehead and see if I have a fever?" Is it easy to get touch this way? Does it feel nurturing, or clinical?

The first question is permission. It's about what the initiator wants, and doesn't really take into consideration what the recipient wants. Oftentimes, when someone wants something from another person, they ask for permission.

The second question is consent. It gives the recipient agency, and takes into consideration their desires first and foremost. It also gives the recipient an opportunity to say no.

If you are going to be asking people if you can touch them, or if they will touch you, consent is highly preferable to permission. Questions that start with "would you like" are the way to go.

As you have seen in this chapter, there are many nuances to touching other people, and many of us don't understand or consider the ramifications of them. Consensual touch is a small way we can make the world kinder and more equitable. In the next chapter, we'll be digging into the philosophy of nurturing human touch.

Amma

On the first Friday of summer, my friend Blair and I woke up at 6:30 a.m. in a hotel room in North Dallas. We quickly dressed, packed, and hopped in the car without having break-fast, and drove another 30 miles north to the Denton Conven-tion Center. Denton is best known as the home of the University of North Texas, an artsy party school, but the Convention Cen-ter and its adjoining Embassy Suites were playing host to Amma (the "hugging saint") for two days and nights. Blair and I had made the trek from Austin to partake of her embrace.

I had heard stories of people waiting for hours to have this experience, but there were only a few hundred people assem-bled. Because it was my first time ever, I got fast-tracked to receive my hug from her, and got in a different line from Blair. One of the helpers gave me a small slip of paper with my num-ber, F1, and I gripped it tightly as I found a seat in the third row.

When I heard it would be another hour until Amma's arrival, I wandered back out into the hallway to get some food from the vendors who had set up shop. After steadying my blood sugar, I deposited my jewelry, phone, and wallet into a bag for safe-keeping. Because thousands of people were going to be held by Amma, we had to make sure we weren't wearing jewelry that would catch on her clothes and rip them. This was the sort of detail I would never have considered, but if my resume included "hugging millions of people," I suspect I would have learned that lesson the hard way.

I returned to my seat and made friends with another first timer, a former Dallas resident who now resided in San Fran-cisco and had been part of a Burning Man camp called Hugzilla. She and I chatted while we waited. There was a commotion as people began lining up by one of the side doors, to be close to

Amma as she came in. She was unremarkable in appearance, a heavyset Indian woman in her 60s wearing a plain white salwar kameez, but people were enthralled, wanting to be close to her. She travels with an entourage of 400 people, most of whom are unpaid.

The room fell silent as she took her seat on a dais on the stage. We closed our eyes, and her right-hand man led us in a meditation. It was a very different experience to be in a large group of people quieting their minds than trying to do it on my own at home.

Once the meditation was over, we were asked to take off our shoes and move to the right side of the room to queue up for our hugs. My new friend and I moved from chair to chair, closer to the front, as those before us received their embraces. We watched as people of different ages, genders, and backgrounds were hugged. Many of them were crying when they came away.

As is befitting a guru who hugs people for hours on end without a break (she can go for nearly 24 hours straight), Amma had an amazingly efficient staff. There were lines on either side of the stage, and her helpers would get the person from the right side of the stage kneeling in position to receive their hug, while the person from the left side was being hugged. Once the hug was done, the person on the left side was pulled out, and the person on the right side was pushed in.

Finally, it was my turn. I kneeled down, and watched her hug the person from the left side of the stage. Her helpers pushed me forward. I had brought with me the picture of my book cover, and held it out for her to bless. She stuck her index finger in a paste that smelled of roses and sandalwood, and touched the image twice. She then wrapped me in her arms, pulled me close, rocked me back and forth, and said "mama mama mama" to me over and over.

It felt wonderful to be held by someone with unconditional love for humanity, an embodiment of the Hindu goddess Shakti, the mother archetype. I sank into her strong arms, and relaxed. I've given many an embrace where I can feel the person I'm hugging melt as they realize they are safe. It was nice to be on the receiving end of a mistress of the art. If anyone understood the power of nurturing human touch, it was this woman.

When the embrace broke, she pushed into my hand a rose petal and a Hershey's kiss and her helpers pulled me back. I staggered over to a chair on the stage to sit and observe. My new friend from San Francisco joined me shortly, and we hugged and cried, feeling overwhelmed by the beauty of the experience. We sat on the stage for close to 30 minutes, quietly observing, marinating in the loving kindness Amma radiated, until finally we were asked to leave to make room for others.

## CHAPTER 7

Philosophy of Nurturing Human Touch

O THER THAN AN occasional hug with friends, most of the touch we receive as adults is sexual or erotic in nature. The other types of touch we frequently receive are unwanted or unconscious, so the concept of nurturing human touch is a foreign one to most people. In this chapter we'll take a look at some of the underlying theories we want to embody when we string those three words together and contemplate sharing touch with other people.

Nurturing

Nurturing human touch isn't as much foreign as *forgotten*: We began our lives with it. While sleep, food, and elimination were important to our development, our parent(s) may have also spent a lot of time holding and touching us. It is how we received comfort, and recognized that we were safe and cared for. If you've been a parent, you've been on the giving end and observed its simple power, but it's likely you haven't contemplated receiving it as an adult, or offering it to a friend or even your partner.

The dictionary definition of nurture is to care for and encourage the growth or development of someone or something. It has the same etymological roots as nourish and nutrition. When scientists look at the question of nature vs. nurture, they are trying to determine whether genetic or biological factors have a greater influence than environmental factors. It's no coincidence that children raised with a great deal of touch and cuddling reap positive benefits from this early caregiving long after they are out of diapers. What might it do for adults?

Nurturing human touch is touch that feeds you and contributes positively to your attitude, disposition, and overall well-being. It can make you feel calm, refreshed, relaxed, and blissful, in much the same way a good night's sleep or a delicious meal shared with friends does.

Presence

When you are present during nurturing human touch, you are physically, mentally, and emotionally available. You are not worrying about your next meeting, or wondering who is texting you – you are paying attention to the people in front of you, and the moment. The mood can be somber or silly...there is no need to change or fix. You don't need to do anything other than be a human who is with other humans.

Nurturing human touch need not be about long-term commitment, or an ongoing relationship: It can last a moment or a lifetime. Compassion is at its core. It tells a different story about humans than the one we normally hear about competition and exclusion, and requires consent, an open mind, and an open heart. It's a *namaste* in non-verbal, physical form.

Belonging

Humans are social animals – we need each other to survive and thrive. When you touch (or are touched) with the intention of being kind and tender and taking care of another person, your actions say, "You matter to me. I care about your well-being, and I want you to know that you are important. I see you not as broken and flawed, but as whole and beautiful, and as a valuable member of my village."

Belonging is a powerful motivator, and is often used in service of division and distrust. But underneath the rhetoric, we are all part of the human family. Nurturing human touch allows us to recognize and demonstrate this truth.

Affective Touch

Affective touch refers specifically to slow, gentle touch that conveys social support and promotes bonding and cooperation. It is different from discriminatory touch, which senses when we touch something that is smooth or rough, hot or cold, large or small. Affective touch allows us to give and receive tangible kindness.

Nurturing human touch conveys kindness, warmth, calmness, compassion, comfort, and support. You can practice feeling different emotions and communicating them through touch. While the receiver might not pinpoint the exact emotions, they will notice that the touch feels good. You can also practice touching people while you aren't paying attention, and see if it feels different to them than when you are paying attention to touching them.

What Nurturing Human Touch Isn't

Massage

Massage has gone from being a shadowy enterprise that people view as a front for prostitution to a billion-dollar industry with Massage Envys on every corner. Massage is now an accepted form of therapy and relaxation, and has been a fantastic way for people to get their touch needs met for the past 20 to 30 years.

So what is the difference between massage and nurturing human touch?

The goal of massage is to manipulate the muscles through pushing, pulling, and prodding. It literally gets under the skin to work on connective tissue and muscles to re-align the body. There are some massage practices that take place on the floor (Thai) or in a swimming pool (Watsu), but most often it is on a table, and the client is unclothed and draped with a sheet. And while it's possible for most of us to give a decent shoulder rub, without proper training you can really fuck someone up if they have an injury. Massage also requires a good deal of physical strength.

The goal of nurturing human touch is to lightly touch the skin without trying to reach the muscles or connective tissue. While skin-to-skin contact is lovely, it's not required; nurturing human touch can be very effective through clothing. It is also more egalitarian: You can go back and forth between being the giver and the receiver, or sometimes you are both at once. It doesn't require any special training or equipment and can be done quickly and easily in almost any setting.

On a neurological level, massage activates the brain's dopamine system, while light touch releases oxytocin. On an emotional level, massage can feel more clinical, while nurturing human touch feels warm and comforting. Massage therapists have loved coming in for Karuna Sessions because they spend a lot of time giving touch. Receiving nurturing human touch allows them to recharge their caregiving batteries.

Both massage and nurturing human touch are wonderful and necessary at different times, for different ailments. Massage therapists often tell me they have clients who need to be held. In the future, nurturing human touch will be understood as distinct and separate from massage, and viewed as a tool for comfort, connection, respite, and relaxation.

Sexual/erotic touch

It's often said that the mind is the most erotic organ of all. Sex often starts in the brain: Texting your lover and building anticipation for your evening rendezvous is a perfect example. Once you are in the same physical space, and the seduction begins, fingers touch skin, and the arousal moves from the mind to the body.

This is different from the intention of nurturing human touch, in both theory and in practice.

Erotic touch is about arousal. Nurturing human touch is about relaxation. It is calming, not exciting. No kissing is involved, and clothes stay on. If you like, you can pretend you're in a Japanese manga comic book where there are no genitals drawn; for the purposes of nurturing human touch, the genitals don't exist.

While our intent is not to create sexual arousal, it does sometimes happen when physical contact occurs. It's not an uncommon occurrence: Billions of people get turned on while walking down the street every day. The difference is whether or not you act on it. If you're feeling aroused, take a few deep breaths, change positions, ask the person who is touching you to stop for a minute. There – don't you feel better?

You might practice nurturing human touch with a lover, but

chances are you will be with people where sex isn't an option. Making comments about someone's appearance, or your desire to jump their bones, crosses a boundary and will destroy the trust you are building. It's also short-sighted: It's unlikely that you will get either sex or nurturing human touch from the person who is now backing away from you. If you want to mix nurturing human touch with sexual touch, that needs to be negotiated ahead of time...and should probably be avoided on your first excursion into this new world.

As adults, most of our touch experience before having children does come from sexual encounters. I find this strange, as in this day and age, sex has been reduced almost entirely to the visual: What a person feels and smells like takes a back seat to their appearance, and many single people experience sex visually by watching others have it. Sex also tends to focus on the performative instead of the pleasurable. While sex can involve a huge range of activities, it is often defined as penis-in-vagina intercourse where the goal is for the male to ejaculate.

By contrast, nurturing human touch invites you to slow down, be playful, and – most importantly – enjoy the experience without having an end goal. This can be a huge relief in a world where we are constantly striving for something different, or *more*.

This chapter has introduced you to some of the values and ideas that go into nurturing human touch, and has differentiated it from other kinds of touch. Next we will talk about the hands-on ways to give and receive touch.

Whitney

Shortly after my breakup, my friend Whitney invited me over to their house for a massage. They had recently transitioned out of their career as a professional massage therapist into full-time parent of two rambunctious kids with an online marketing business on the side.

Whitney is a believer in the power of human touch. In addition to their experience as a massage therapist and snuggling with their kids every night at bedtime, they were an early guinea pig for Karuna Sessions. After a session, they would often comment on how nice it was to be touched by someone who didn't want or need something from them.

I arrived right around bedtime to see the kids before they went to bed. This proved to be a bad move: The older one was so excited to see me that they kept getting up and interrupting our massage. Whitney put them back to bed several times, and around the fifth time they stayed in their room and finally went to sleep.

Whitney and I hadn't talked to each other in a while, and spent most of the massage catching each other up on our respective lives. I talked about re-adjusting to life on my own; they talked about the difficulty of being constantly present, engaged, and focused on two young children. When I suggested that we snuggle post-massage so I could take care of them, they eagerly accepted.

They finished pummeling my muscles and I got up, dressed and drank a couple glasses of water. Recently they had acquired a massive brown microfiber bean bag chair that took up most of the corner of the room, and we climbed into it. They leaned back against me, and I wrapped my arms around them. It felt good to be close to someone I trusted, and to be taking care of them instead of endlessly processing my sadness. For a few moments, we were both silent. Our breathing slowed, and our muscles relaxed.

Whitney sighed deeply, and looked up at me. "I'm really glad to have you in my life," they said. "Back atcha, sugar," I told them. We continued to cuddle, talking, and giggling like teenagers, having some much-needed downtime. After about thirty minutes we both began to get sleepy, and called it a night.

CHAPTER 8

Touch Basics

JUST AS A baby learns words before understanding grammar, it's important to know the *what* of touch before learning the *how*. In this chapter I'd like to cover some basic information about touch techniques.

Nurturing Human Touch in Practice

Don't overthink it

There is no right or wrong way to touch another person. If it's consensual, and they are enjoying it, you are doing it right! For our purposes, though, there are two prime directives: Touch slowly, and touch lightly.

The best way to get good at touching another person is to tune in to your own experience. Am I applying too much pressure? Is there a rhythm to what I do? Am I enjoying it? This will likely take some practice, as it will be a new thing, especially if you live in your head a lot. Fortunately, it's hard to physically injure somebody while you're learning, though you will often find that people will cry tears of relief or laugh uncontrollably when they haven't been touched with tenderness and compassion in a long time. Sometimes they will cry and laugh at once.

Touching vs. Being Touched

Our fingers have nerve endings in them that are designed to register tactile sensations and give feedback. Similarly, our skin has nerve endings that specifically respond to touch.

If you are engaged in an activity where you are the giver, you should be open to feedback from the person you are working on/with. You will want to know if you are going too fast or too slow, if the receiver wants a different type of touch, or if they would prefer that hands go elsewhere. But if the feedback you are getting is "That feels great!" then the best thing you can do is be quiet and focus on enjoying your own experience.

If you are the receiver, ask yourself two questions: "Does this feel good?" If the answer is no, then ask "What *would* feel good?"

If the answer to the first question is yes, then your best course of action is to relax and enjoy. This is harder than you think. I've seen so many clients who want to give back, interact, and reciprocate when we are touching them. Don't you have enough situations in your life where you are required to be doing things? Can you *allow* yourself to be cared for and nurtured, to let someone else handle the details for a few minutes?

When your desire as a receiver is to give back, you are thinking about the other person and what they want and need, instead of allowing your own needs to be met. Take a vacation from thinking. Breathe, and enjoy being in your body for a moment. Trust me – there will be plenty of people clamoring for your attention and care when you go back into the default world.

Nurturing Human Touch Techniques

Below we will look at some specific ways to hold and move your hands and fingers on the human body. Some of them may be obvious, but some of them may be new to you. Either way, it's good to have a variety in your repertoire.

Still, open, full hand

Open your hand and place it palmside down on the person you are touching. Pressure should be light to medium. Your fingers can be slightly splayed to cover more surface area, and may curl a bit if they are on a forearm, shoulder, or lower leg.

Still, open, full hand is a great way to start: It reassures the person that you are present, and gives them a moment to adjust to having some human contact. When you have been moving your hands and fingers and covering a lot of territory, it can be helpful to pause for a moment here and there to let the other person come to a place of stillness and integration.

Moving, open, full hand

From the still, open, full hand position, you can start moving your hand along a person's back, arms, legs, or torso. Because you are using the whole surface of your hand, you can activate a lot of nerve endings at once. Experiment with moving your hand back and forth, in long strokes or in circles – circular strokes feel wonderful on the belly. If you have your hand on their face, short strokes are reminiscent of being tucked into bed by their parent.

Moving finger pads

You will be using the pads of your fingers and thumbs, and moving them lightly across the body without breaking contact. While there is less contact between you and the other person than with a full hand, this is a case where less is more. This is a delicate touch that makes the receiver more aware of the contact. It reassures them that the giver is being careful not to overwhelm the receiver, and is paying attention to potential physical injuries.

Drumming fingertips

Again using the pads of the fingers, place the fingers one at a time on the person's body, in the same fashion you would strike a keyboard. This is better used on the body than the face, and may or may not feel pleasant for the receiver. It is playful and fun to do for the giver.

Cat scratches

These involve using the tips of the fingers and the fingernails. Gently drag the nails across the body. Cat scratches provide a bit more focused contact. We frequently use them during our sessions on backs and stomachs and arms. It also gets the blood flowing into all the little veins and capillaries of the skin, delivering more oxytocin-y goodness.

Hands encircling

You can use both of your hands to hold another person's hands, arms, feet, or ankles. While you don't want to squeeze, you do want to have contact with most of your palm, fingers, and thumb. You will want to keep your hands still.

Hugging

When you get on the nurturing human touch welcome wagon, you will probably find yourself giving – and receiving – a lot of hugs. If you have a well-established friendship or relationship, you are probably already hugging the people you know hello and goodbye when you see them, but for the purpose of our education, we have protocols. These protocols will come in handy if you have opportunities to hug strangers.

Don't just reach out and grab for the other person: Always ask, "Would you like a hug?" and wait for their answer. If you get an affirmative answer, let the person decide how distant they are from you and how long the hug lasts. This might mean that the person wants a one-second, A-frame hug (where

---

TRY ME! Let your fingers do the talking

Practice the different sorts of touch described in this chapter. If you have a friend who will feel comfortable being your guinea pig, you can do them on another person. Ask for feedback: What feels good, and what doesn't? Are there certain body parts that prefer one type of touch over another?

If you don't have someone you can practice on, you can do them on yourself.

---

feet and bodies are apart and you lean in toward each other, mainly touching shoulders) because it might be all the contact and closeness they can handle. You can hold the hug until the other person pulls away.

Tickling

As a giver, you want to avoid tickling your receiver, as it will cause the body to tense up. This is not relaxing, and it may hark back to older brothers and sisters picking on their younger siblings. It's a good idea prior to any kind of touch experience to ask the person if they have any ticklish spots so you can do your damnedest to avoid stimulating them.

Sometimes a receiver will find that something tickles because they are feeling over stimulated. They might also find a particular spot ticklish because their body is being protective of that area due to a past injury or memory. If this happens, move to a open, still hand while the receiver breathes deeply before continuing moving your hand.

Learning about different types of touch is fun, both as the giver and the receiver. Next we will talk about the theoretical framework necessary to design your nurturing human touch relationships.

Sasha and Taylor

I'm in the kitchen when I hear the knock on the door. I run to open it, squeal, and hug each of the people who are standing there.

Sasha and Taylor have been away for the summer, and I have missed them. I invited them over for dinner, catching up, and snuggling. We've known each other for less than a year, but the three of us can usually be found at parties, arms around each other in the corner and talking intensely, trying to make up for all the years we missed before we met.

We enjoy some soup and salad, and they tell me about their adventures. Taylor spent the summer in Vermont, and road-tripped back to Austin on their new motorcycle. Sasha spent two months traveling in Costa Rica and swam in oceans, rivers, and hot springs. I catch them up on my breakup and the progress with my book.

As dinner winds down, I ask each of them what they want from the snuggle portion of the evening. Taylor says they would like to lie on the floor with their head in Sasha's lap while Sasha and I place our hands on their arms, legs, and torso. They want our hands to be still, except when we move our hands to a different spot. Sasha wants to lie down with Taylor and I on either side of them and have us lightly brush our fingertips rapidly across their entire body from head to toe. I tell them I would like to end our evening with the three of us spooning, and each of us taking a turn being in between the other two.

I just bought a fancy modern futon couch, and I lower it down flat so we don't have to be on the floor. I put on a playlist of ambient music that a friend made me for the Snuggle Salon, and prepare to share touch.

We start with Taylor's request. They lay down on their back with their head in Sasha's lap. Sasha places one hand on Taylor's forehead, and the other on their heart. I move to the other end of the couch, and wrap my hands around their feet. Our conversation ceases. I can feel my breathing slow down. Sasha, who normally can't sit still, becomes quiet, and looks down at Taylor tenderly. Taylor's face relaxes. Sasha looks up – our eyes meet and we smile at each other. Sasha moves her hands onto Taylor's forearms, and I move mine onto their shins. Sasha and I move our hands to a different position every few minutes until we have covered most of Taylor's arms, legs, head, and torso. We withdraw our hands, and sit quietly. Taylor opens their eyes, looks up at Sasha, and then looks at me. "Thank you both," they say. "I finally feel like I'm off my motorcycle and have arrived in Austin."

The three of us get up to stretch and drink some water. Sasha lies down on their back. Taylor and I sit on either side of them, and begin moving our fingers from the top of their head to their feet without breaking contact. We start varying speed and direction. Sometimes we use our nails; other times we drum our fingers. Sasha begins to giggle. "That feels delicious!" they say. Soon Taylor and I are laughing as well. While Taylor wanted to be grounded, Sasha wants some levity, and the mood becomes festive. After 10 minutes, Taylor and I place our hands on Sasha's hands and shoulders to signal that we are done, and then withdraw them altogether. Sasha sits up, and hugs each of us.

Since Sasha is already in between us, I suggest that they turn on their side. Taylor and I lie down on either side of them. Taylor and Sasha face each other, and I spoon Sasha's back. The three of us begin to adjust our arms to find comfortable positions for them. Once we are arranged, we stop talking. Within a minute, all three of us have synchronized our breathing. The room feels peaceful. I move my hand up to Sasha's shoulder, and begin gently stroking their arm. "I wish I could fall asleep like this every night," Sasha murmurs. After a few more minutes, Sasha squeezes each of our hands, thanks us for holding them, and trades places with me. As Sasha and Taylor move closer to me, I breathe deeply, and shut my eyes. I put my head on Taylor's chest, and listen to their heartbeat. I feel like time has stopped, and that the three of us are adrift in a boat. My mind stops its chattering, and my heart begins to smile.

The end of a song reminds me where I am. I become aware of the room again, and tell my friends we should switch places. This time, Taylor lies in between Sasha and me. Sasha spoons Taylor's back, and Taylor and I face each other. When we all find comfortable positions, we quiet down. After a couple of minutes, I feel Taylor trembling a bit, and realize that they are crying. "I haven't been held like this since I broke up with my boyfriend last year," they say. "I miss this so much." I bring my hand up to Taylor's face, and wipe away their tears. Sasha and I continue to hold Taylor until their sobs subside and their breathing deepens again. We lie together, quietly, until Taylor pulls away from me. "Thank you," they say. "It's been a long time since I cried like that."

I get up and put on the kettle to boil water, and make us a pot of tea. The three of us sit quietly, sipping our tea, each of us lost in our thoughts. I'm struggling to keep my eyes open; I know I will sleep well tonight. I wait until Sasha and Taylor are finished with their tea and tell them I need to get to bed. "I'm sleepy too," Taylor says. "I'm glad I didn't ride my bike over here tonight."

I walk them both to the door, and the three of us put our arms around each other for a hug, our foreheads touching. "I love both of you so much!" Sasha exclaims. "Let's do this again next week!" "Definitely!" says Taylor. "This was the best night I've had in ages." They walk out the door and into the night, leaving me smiling in the doorway.

## CHAPTER 9

Designing Nurturing Human Touch Relationships

I N THIS CHAPTER we are going to look at the conceptual framework you need to add nurturing human touch to your existing friendships. Because most of us equate touch with sex, it will be important to approach both the idea and the physical action of touch carefully, thoughtfully, and deliberately. Since we want to enhance and improve on our relationships, not destroy them, we will need to design what we do instead of letting it happen organically.

Archetypes

An archetype is defined as "a very typical example of a person or thing." While the concept of the archetype is most commonly used in psychology and mythology, all of us inhabit multiple archetypal roles every day. You may not realize what's happening, but once you become aware of the roles you take on, you can start viewing all your interactions from this perspective.

A few of the archetypal relationships you may engage in daily from either side of the equation:

Boss/employee
Customer/cashier
Parent/child
Beauty/beholder
Barista/uncaffeinated zombie
Performer/audience member
Driver/pedestrian
Teacher/student

These relationships are common, and have well-known scripts and roles. When you attend a play, for instance, you sit down in your assigned seat, turn off your cell phone, become attentive when the curtain comes down, refrain from whispering to your neighbors, and clap at the end.

The performers, on the other hand, are putting on makeup and getting dressed, limbering up and calming themselves, waiting on the side of the stage for their cue to enter, walking on stage, reciting their lines, and walking off when their part is over.

While a few of the particulars have changed over the centuries and in various cultures – Shakespeare didn't have to deal with self-absorbed people Tweeting during a performance – most people have been in these roles often enough to make them second nature.

Think back, though, to the first time you attended a play. You had no idea what to do or how to behave, and looked to your fellow audience members for cues on what you should or shouldn't do. Because you were having an adult experience, you wanted to make sure you got it right so that you could do it again.

When you start looking at your interactions through an archetypal lens, it can make difficult situations less emotional and personal. This is super helpful when there's a new clerk at the grocery store who is being trained and taking a million years to scan your groceries when all you want to do is get home.

It also makes it easier to step into new roles that you might not already know...like, say, a person who wants to receive and give nurturing human touch. You can see these new roles as one aspect or facet of your being, as opposed to having them be your entire identity, or something that defines you.

The nurturing human touch relationships you are going to be creating have no familiar archetypal roles or scripts: This is a brand-spanking-new way of relating to others. There is some overlap with cuddle parties or BDSM scenes, but those are considered "alternative" and haven't yet gone mainstream (*50 Shades of Abuse Disguised as a Romance* notwithstanding). Because nurturing human touch is scriptless, you will want to be careful to practice as you would for a performance: Learn your dialogue, block, and then rehearse, rehearse, rehearse.

I will be the director for your performance, but I promise not to be the eccentric old codger who screams and throws her script on the floor because the actors just can't understand her vision! (I do, however, reserve the right to wear a beret, chain smoke, and pace nervously.)

The Four S's

There are four guiding concepts you will use to design your nurturing human touch relationships. They are *Slow*, *Structured*, *Specific*, and *Safe*.

Slow process

Life in the 21st century moves at a rapid pace. While once you would have waited months to get a response to the letter that traveled across the ocean via boat, today you think someone is ghosting you if they don't answer your text message within two minutes. Gone are the days of dial-up modems taking five minutes to load a rudimentary web page. And while the slow food movement encourages you to savor every bite, chances are you gobbled down a sandwich for lunch today while simultaneously checking your Facebook feed, listening to a podcast on productivity, and texting a friend about dinner plans.

As noted, you will be creating a new form of archetypal relationship that does not have a script. There will be confusion and uncertainty. While you may read this book and think to yourself "I want to be in a giant puppy pile with all my friends!" another of your friends may be thinking "This all sounds really weird, but I'd be willing to try a hand caress." Do not make the assumption that everybody is on the same page, because it's highly likely that they are not.

Going slowly will be important in the beginning. You will want to do some sleuthing prior to approaching your friends about adding touch to your relationship, and giving them plenty of time to respond. You may also want to break down your experience into several shorter encounters so that people will be able to reflect and integrate. We will talk about this process at length in Part 2.

When you are going to ask someone to touch you, or ask if they would like you to touch them, give them time to consider what they want before they answer. Many of you won't know what you want, or what you like when you start out, and that's okay. It's rare that we are asked to think about what we want and voice our desires.

Allowing for a lot of physical space between people is also helpful: Starting with the hands or feet (as we will be doing) lets you get comfortable with the idea of having physical contact and expanding your relationship to a different dimension. Everyone will bring with them differing levels of comfort with touch, unspoken needs and desires, and a lifetime of good and bad experiences.

Nurturing human touch is powerful. Many of the people in

your group will be dealing with feelings of shame and unworthiness if it's been a long time since someone touched them with compassion and kindness. They may think they don't deserve it. They may fight it. They may feel angry, or shut down emotionally. There is no right or wrong way to react. As the hippies say, it's all good.

It may be that you try the first set of activities and find that you need time to process and integrate the experience. You may not come back to nurturing human touch for a couple of weeks or couple of months. Hell, you may throw this book across the room after you read it and get pissed off because you think this will never happen for you. Your goal is to get to a place where the people who share touch with you are saying YES to this experience. Go slowly; it will be worth it!

Structured experiences

When you are adding nurturing human touch to a relationship that has not previously had a physical dimension, it is important to be overly, annoyingly formal about it. As in architecture, form will follow function, and the function here is to allow people to get physically closer to each other.

Make sure people know your intentions, and that they have consented to participating in an experience that involves physical contact. Pick out music, and plan snacks or a meal for after you are done. Set up the space in advance. Have a start and finish time. State clearly at the beginning: "It's time for nurturing human touch." Or make up your own phrase – mine is "The otters are now in the hot tub."

Ask questions that will allow people to talk about their experiences with touch in the past, and run through the boundaries exercises. (All of these things will be discussed in greater detail in Part 2.) When you are finished, ask people to talk about their experiences, and then state clearly: "Our time for sharing touch is over." It's like doing a BDSM scene, but touching and cuddling replace ropes and whips.

This might seem overly cautious, but it serves a purpose. To go back to the theater analogy, you want to be rehearsing the play "We Created Nurturing Human Touch Relationships Last Night," not the musical "My Friend Gave Me a Hug and I Think They Might Be Coming On To Me." You want to make sure that everyone is reading the same script and knows how the story unfolds. Assumptions and ambiguity are not your friends.

To further ensure that things don't get weird, this book is designed to establish nurturing human touch relationships in a group setting and rebrand touch as a social activity. Because the vast majority of romantic relationships involve two people, it's much easier to misconstrue intention and gestures if touch is shared one-on-one. Being in a group might be weird and awkward, but it won't be weird and awkward because your friend thinks you want to boink them. Primates often share touch in groups, and since we are merely monkeys with car keys, it might feel familiar after you get out of your head and into your body.

Once you have gotten comfortable with touching your friends and have a mutual understanding around requests for it, it will be easy to plan an evening of snuggling on the couch to watch a movie one-on-one. Because you learned about nurturing human touch together, your friend won't be left wondering if your request for snuggle time is actually a ruse for trying to get in their pants.

Specific touch

As we discussed in Chapter 6, touch needs to be consensual to be pleasurable, and many of us receive uncomfortable – and unwelcome – touch on an all-too-regular basis. If we are creating a new paradigm of nurturing human touch relationships, shouldn't everyone get the opportunity to receive exactly what they want? Wouldn't *you* like to be touched in ways that feel good to you? The best way for this to happen is to be specific about it.

This approach flies in the face of one of our biggest romantic myths: that your soulmate will know exactly how to touch you to bring you to the heights of ecstasy without any direction. This concept has led to billions of dissatisfying sexual encounters, faked orgasms, and hurt feelings. Unfortunately, the opposite scenario, giving direction, is considered unromantic.

When learning about nurturing human touch, you want to be unromantic. Like, *hella* unromantic. You don't want to perpetuate this romantic myth and go home feeling unhappy because you didn't get what you wanted and needed.

Of course, as you start out, you will likely have no idea what it is you want – hence a set of specific activities to do. The idea that you can pick and choose what sort of touch you would like to receive might even break your brain a little bit. It will take a bit of time and experimentation, both to figure out what feels good and to learn how to ask for it. Be willing to say no, and offer feedback when requested. After a bit of practice, it will get easier.

Safe environments

Safety starts in the body.

Our species began, and survived, by living in close physical proximity to one another. Babies may be able to survive without adults who will hold them, but they won't thrive. *The feeling of safety is imprinted onto our bodies via our earliest experiences of touch, and thus it is vital for said touch to feel nurturing.*

You and your friends are arriving to this place in your life with millions of experiences that have shaped your beliefs and perceptions. You have been loved and hurt, praised and shamed, bullied and cherished, admired and judged, pleasured and attacked. Whether your life has been easy or hard, your emotions reside in your body. This is especially true with trauma.

When trauma is stuck in your body, the world feels like a battle zone where you constantly have to be on high alert and you cannot trust anyone. Muscles tense; you look for danger and expect things to go wrong. Not only does this feel horrible, but it affects your physical and mental health, and makes it difficult to create and maintain relationships. It can also do a fair amount of damage to your pocketbook.

In 2004, I had my psychological boundaries aggressively crossed for the amusement of others. The experience left me traumatized, and physically ill for a year. It happened in a very public place with hundreds of people around, and after the fact, people refused to believe I had been injured. For nearly a year, I tried to get them to understand, and found myself constantly on the defensive while trying to get my health back. People I thought were my friends attacked me and blamed me for my experience. I finally realized that they didn't want to understand, and didn't care that I had been hurt, and I decided to walk away from the group altogether to prevent having the wound continually ripped open.

While this experience happened over 10 years ago, anytime I find myself in a place where I'm around these people, my body says, "Nope. This is not a safe place for you. GET OUT. NOW." While my rational mind knows that I'm not in any physical danger, I have yet to find a logical perspective that will convince my body otherwise.

You may have a story like mine. Some of the people in your friend circle probably have stories much worse than mine. Your success with nurturing human touch depends on ensuring that everyone feels that their no's will be honored, and that they don't have to worry about being violated.

There are many things you can do to create an environment

that feels safe to participants. These include:

* Modeling clear communication, compassion, consent, and strong boundaries.
* Getting explicit consent before you touch anyone, even (especially) if your intention is to demonstrate something.
* Making sure participants get plenty of practice with the boundaries exercises, and that they feel good about saying NO.
* Giving a thorough explanation of what is going to happen before each activity that involves touch, including how long each segment will last, and asking people if they want/need to opt out.
* Allowing people to decline to participate without them feeling pressured to say yes or having to answer questions.
* Asking if they have questions or concerns after each activity.
* Checking in frequently to make sure that participants feel emotionally and physically comfortable as you are going through the activities.
* Letting them know that if anything comes up for them after they leave that they can/should contact you. You may want to proactively call/text/email them later in the evening or the following day to check in with them. We will discuss this at length in Chapter 14.

Nurturing human touch gives you an opportunity to write a new story about people, one that involves care, trust, compassion, and kindness. It also allows your body to experience a feeling of being safe and relaxed. Wouldn't it be wonderful to create a sanctuary where you and your friends can let down your guards for a little while? As adults, we have so few opportunities to be carefree.

TRY ME! Play the Part

Spend a day paying attention to the different archetypal roles you inhabit. Were there some that felt more natural to you than others? Were there qualities that were essentially "you" that you brought into every encounter? Did you put on a mask when a role felt uncomfortable, or when you were required to do things you didn't enjoy? Did you encounter people who were aware of their roles? Were you able to detach a bit in difficult situations when you realized that you were playing a role?

You have just completed a crash course in touch and its role in our bodies, relationships, and culture, and now have the theoretical framework to share it consciously. In Part 2 of this book, I will walk you through each step you will need to take to share nurturing human touch with your friends.

Part II

The Step-by-Step Guide

Top 10 Reasons to Share Touch with Your Friends

If you were alive 1,000 years ago, you wouldn't have given a second thought to sharing touch with the people in your circle – it would have been part of your everyday life, regardless of your sexual/familial relationships. And while sharing touch with friends is a simple, elegant solution for getting your touch needs met, it is likely to give you pause. Below you will find 10 compelling reasons to give and receive touch with your friends.

Number 10:
You care about them.

One of the most frustrating habits of modern daters is ghosting. How can someone have sex with another person and yet not be kind enough to say, "I'm sorry, but this isn't working for me"? Your friends may do or say things that hurt you, but you know that you love them, they love you, and ultimately you want the best for each other.

Number 9:
You support each other.

When you are just getting to know someone, you tend to not talk about the hard stuff you are facing, for fear of appearing too needy. Your friends, on the other hand, know about your domineering boss, your struggles with depression, your last breakup, and your aunt's cancer. You share the good stuff and the bad because you know your friend has your back, and will continue to be there when life isn't coming up rainbows and unicorns.

Number 8:

You enjoy spending time together.

Frequently, you meet and bond with your friends during shared activities. If you have known someone for a while, chances are you have already played video games, gone grocery shopping, rehearsed a few songs, helped them move, or gone hiking. Whether it's trying something new or taking care of the mundane, you are happy to be in their company.

Number 7:

You know them.

When you go out on a couple of dates with someone, you don't know much about their character. Did their last partner leave them because they were physically abusive? Are they asking you about your 401K because there's a string of bankruptcies and foreclosures in their past? With people who are your friends, you have already established a level of trust and transparency that is conducive to sharing nurturing human touch.

Number 6:

You like them.

Many people have sex with people they don't like because they feel they don't have suitable options. We are encouraged to say "Fuck yeah!" when it comes to the people we date, but often our response is "I guess they will do until something better comes along." Your friends are the people you say "Fuck yeah!" about already.

Number 5:

You feel safe with them.

Because the goal of nurturing human touch is to relax your body, it's imperative that the people you give it to and receive it from are trustworthy. Your friends are already in your inner circle. Who better to feel safe with than someone who you have talked, traveled, and partied with?

Number 4:
You have open communication.

If you've been friends with someone for more than a year or two, you have already been through a few disagreements and conflicts, and have said things that are hard to say, and heard things that are hard to hear. Because nurturing human touch requires us to voice our desires, it's nice to indulge in it with people who will actually listen to what you are saying to them.

Number 3:
You feel comfortable around them.

When you are getting to know a potential romantic/sexual partner, you are on your best behavior. You excuse yourself if you have to burp or fart, take extra care with your grooming, make sure you don't discuss past lovers, and skip the dessert because sugar makes you bloated. With an established friendship, you don't need to worry about making a good impression or editing yourself.

Number 2:
You have established intimacy.

Nurturing human touch requires vulnerability and tenderness. You will want to share it with people who have seen you at your best *and* your worst. It will be easier to let your guard down with someone who has witnessed your fears, your dreams, and your celebrations or who has held your hair back while you're puking after a long night of drinking.

Number 1:

You want to know them better.

One of the best ways to deepen your relationships is to share an experience that allows you to (a) explore different parts of your psyche, (b) collaborate on creating something or reaching a goal, (c) learn something new together, and (d) talk about how your perspectives and assumptions have changed. Nurturing human touch allows you to do all of these things for free, without even leaving your living room, with people you adore. #winning

Chapter 10

Assembling Your Crew

Admitting to your friends that you want to share nurturing human touch with them will be one of the hardest things you ever do.

I suspect that if you're reading this book, you are a rebel, a trailblazer who likes to be in the vanguard of change. You're an early adopter, joining communities, networks, and movements long before they become popular. While you might surround yourself with similar people, oftentimes your friends are scratching their heads and asking "Why are they drinking a beverage made from slime mold?"

And then, 18 months later, they are talking about the amazing kombucha flight they tried at that new raw foods restaurant.

The difference with this, though, is that it's a trend you can't follow by yourself: You will need other people for it.

Fact: People are resistant to the idea that they might need human touch.

Theory: People are resistant to this idea because they need it so much.

Remember that when you approach this, you are going up against many cultural myths. We learned about them earlier, but don't underestimate how people will cling to them. Shame, fear of pleasure, rugged individualism, the soulmate myth: These ideas keep us apart. Being kind to and supportive of each other is the proverbial mosquito in the tent of the American dream.

So yeah. It's not going to be easy. Your friends may think you're weird, or ridicule you. They might distance themselves from you. They might get angry with you. They will give you lots of opportunities to hear the word NO.

Give them time and space. (Maybe buy them a copy of this book?) You are asking them to consider rejecting the dominant narrative of our world. Most people like to be followers, not leaders. And while this new story you're telling them – the one where adults comfort and care for each other through touch – has deep evolutionary roots, it's a story they have probably never heard. They will have to listen carefully to hear its whispers.

Likely most of your friends live in their heads, and on their smartphones. If you're lucky, they are somewhat in tune with their bodies. If you're really lucky, they will be able to hear that tiny voice inside of them that says "Touch me." It probably sounds like Janet in *Rocky Horror Picture Show*, but that's not necessarily a bad thing.

One of the biggest drawbacks to being single is that your life lacks nurturing human touch. Not sex, but touch. If you're reading this, it's likely that you would much rather get that need met through your friends, people you trust, instead of going to a touch professional or having hookup sex.

When you approach your friends, though, you are going to have to be vulnerable. So how should you go about making contact?

Asking vs. Guessing

A few years back I came across an article[1] about Askers vs. Guessers, and as promised, it did change the way I interacted with others.

The premise is that there are two types of people: Askers and Guessers. An Asker will ask someone for something they want or need and know that the person being asked will say yes or no. A Guesser will *only* ask for something if they believe that the person being asked will say yes, presumably to avoid rejection. (It gets really tricky when an Asker asks something of a Guesser...the Guesser thinks to themselves "they wouldn't be asking me that unless they thought I would say 'yes'" and might consent to something they don't want.)

A lot of Guessers are touch-situational in that they often read social situations and take into account many factors; they don't want to inconvenience the other person or touch someone where it's inappropriate. Unfortunately, being a Guesser will keep you in your comfort zone. For the purposes of getting your touch needs met, you're going to have to be an Asker.

For your initial approach, though, I will recommend you do a bit of guessing prior to asking when you are looking for people who are already interested in having more touch in their lives. While many of us are touch-deprived, few will admit it. Your mission is to look for early adopters, people who are open to this idea. A few ways you can gauge interest:

Listen to what your friends are saying. If you're anything like me, you constantly hear stories from your friends about how depressing dating is. That person who told you about their hookup where the guy rolled over and fell asleep right after they had sex and they had to quietly let themselves out? They are a good candidate. The person who tells you they have sent out 25 thoughtful messages on a dating website in the past six months, and received zero responses? Yep, them too. These are folks who already know that our current paradigm of love, sex, and relationships isn't working for them. They are likely not getting their touch needs met, and may be ripe for hearing another perspective and trying something different.

Send out some test balloons. Post an article on the health benefits of touch to your social media feeds, and see who responds. When you tweet out "I could really use a long hug today," who makes an offer? A lot of folks are already cognizant of the fact that they are lacking in touch, but they are afraid to admit it for appearing to be someone who doesn't have it all together. They may talk about it if you start the conversation first.

Start asking people for hugs when you see them. Many people hug automatically when they see their friends – it has become a standard hello and goodbye. Ask "Would you like a hug?" and let the other person pull away first. Take note of the people who enjoy hugging and do it really well. Touch enthusiasts are fantastic to have on your crew, and they will benefit from learning how to share touch more consensually.

Have conversations about touch with people in person. Talk about how you feel touch is missing from your life. Maybe discuss a time when somebody gave you a really great hug, and how you slept better that night than you had in months, or how you miss your kids curling up in your lap.

After you have completed your social spelunking, send out an email to a select group of friends outlining what you want. A sample email could look something like this:

"Dear _____, I have been feeling lonely of late. I have many wonderful friends who support me (like YOU), but something's still missing. I have realized that what's missing is nurturing human touch.

"I have noticed that every time I get a good, long hug, I

smile for the rest of the day, and sleep well that night. And I remember that rolling around with my friends as a kid was fun, playful, and relaxing. I could use more of those things in my life.

"I could pay money to go to a cuddle party, but it makes more sense to me to get my touch needs met by people I already know and love and trust: my friends. I am proposing getting a group of four to eight people together to do a little experimentation around human touch.

"I know this is an unorthodox request. Please be assured I am NOT interested in anything sexual: Clothes will stay on and this will not occur in a bedroom. I am interested in creating a new dimension to our existing relationship in a way that feels safe and honors everyone's boundaries. I have been reading a book that gives me step-by-step instructions on how to do this, and I would like to give it a try. You are more than welcome to read the book if you like!

"I am emailing you because I believe that you are someone who understands the importance of nurturing human touch, and would be open to this idea. If you choose not to do this, I will absolutely respect your decision, and will continue to love and support you as a friend.

"I hope you are willing to take a chance to try something new. Please let me know if I can answer any questions."

You will want to tailor your emails for each individual person; you're asking for something intimate, and a form letter might feel impersonal. Feel free to use this one as a jumping-off point, modify it, or write your own.

It may also be possible that some individuals will respond better to a face-to-face query, but it may be harder for you to articulate this very vulnerable request. It may also put your friend in the uncomfortable position of feeling like they need to give you an immediate answer. Other factors, like how long you've known someone, how intimate your friendship is, or if there is prior physical attraction/chemistry, may also be taken into account. As with many things, it's complicated....

It may take a bit of time to assemble your crew. Give people the space to consider this request without pressure. Likely your friends are reflective and thoughtful; allow them time to evaluate all the possibilities. And remember that they are going to have to start working through all of the social conditioning and pressures that you have already considered and begun to work through yourself.

You now have your intrepid crew assembled to journey into this new territory. What items will you need to make your excursion a wonderful experience? Let's take a look at that in the next chapter.

## CHAPTER 11

Preparing for Your First Encounter

A s WE DISCUSSED in Chapter 8, your success in keeping things from getting too strange will hinge on you creating an environment where there is no ambiguity around your intentions. You need everyone to be on the same page when you begin your journey. Once everyone understands what you are doing (and, more importantly, NOT doing) and feels comfortable with it, you can be a lot less formal, but for the time being, treating these encounters like a workshop will be important.

Your Social Contract

We make legal agreements when we download software, or buy a house. We often sign contracts with our employers so we know what they will pay us, when our insurance starts, and how much vacation time we get every year. Yet people recoil at the idea of creating relationship contracts, despite the fact that they can help their partner(s) identify what they want and need – and what is off-limits – in their relationship.

One of my friend circles has been going on an annual campout for 20 years. Between 75 and 150 people attend, and there is a lot of frivolity, some of which involves sex and drugs.

Many years ago, a subset of our group worked out some basic guidelines that allow for a great deal of freedom, while outlining what behavior is considered boundary-crossing and might get you ejected. Our four guiding principles are Respect, Reciprocity, Consent, and Coherence, and they have served us well: Rarely is someone told they can't come back.

But what should be in such a contract, you may be asking yourself? Don't worry your pretty little head, my friend, I've taken the liberty of writing one up. Of course you are free to add things in, or take out points that don't feel right to you, but please send out something along these lines and get an email back from each of your participants that says "I agree."

NURTURING HUMAN TOUCH CONTRACT

Greetings my dear friend _____:
I'm excited about our upcoming foray into the world of nurturing human touch. As this is completely new for everyone involved, I have prepared a contract to help us navigate this territory. You will find it below.

—◇◇—◇◇—◇◇—

I, _____, *have agreed to participate in a gathering to play with giving and receiving human touch. I solemnly promise that:*

* *I will remain fully clothed at all times.*
* *I will not engage in kissing or genital touching.*
* *I will take the time to check in with myself and will say NO if I don't feel comfortable with an activity.*
* *I will allow myself and others to be nervous, awkward, and make mistakes.*

✳ *I will be honest and open about my experiences, and give feedback in the moment when something doesn't feel good.*

✳ *I will not indulge in drugs or alcohol prior to and during our time together.*

✳ *I will allow myself and others to laugh, cry, and – most importantly – have fun!*

✳ *I will give myself good aftercare and not make any life-altering decisions for at least two days.*

If you agree to the terms of this contract, please reply with "I agree." I look forward to seeing you next week.

You may also want to read the contract aloud at your gathering prior to starting and ask people again for a verbal agreement.

Hygiene

You, or some of your friends, might be squicked out by the idea of close physical proximity to other human beings. Gently remind your friends to arrive freshly showered with their teeth brushed. If you have friends who are sensitive to chemicals, suggest that people don't use lotions or perfumes prior to your encounter.

What You Will Need

✳ A private space with access to a bathroom. If you have access to a martial arts or dance studio, or a workshop space, it's nice to have a formal and neutral territory, but it's not necessary. It's possible that you might be in a shared living situation where you don't have access to a common area, or that you have housemates who are not participating, and you need to be in a bedroom. If possible, avoid that. Being in a bedroom with someone other than your child, pet, or longtime S.O. feels like a sexual encounter to most people.

✳ Blankets and pillows and yoga mats. I like to wait until everyone is there to set up the space as it allows people to move around a bit and participate instead of having to sit around and wait and feel awkward.

✳ Print-outs with the questions and activities (you, the "leader," should have this). Alternatively, having this book readily accessible on a tablet or computer will work too.

✳ Some chill music to play during the activities.

✳ A massage table (optional).

✳ A timer of some sort (most smartphones have 'em, but don't forget to put it in airplane mode so as not to avoid beeps from message notifications).

✳ Hand sanitizer.

✳ Snacks and water.

✳ Breath mints.

What Participants Should Bring

✳ Yoga mats, blankets, and pillows.

✳ Comfortable attire (pajamas, yoga clothes, or sweats and t-shirts).

✳ Socks if it's winter (shoes and socks will be off, otherwise).

✳ Water and snacks.

✳ 4 to 6 hours.

✳ An open mind.

Accommodating Those with Special Needs

While many of us struggle with finding suitable people to date, it can be especially hard if you have physical or mental conditions that mainstream society deems undesirable. This can lead to even greater touch deprivation, or touch primarily from medical professionals who poke and prod. If this describes you or some of your friends, it's possible that one or more members of your group will require some extra accommodations.

For people with physical limitations, you can and should discuss these things with them while you are preparing for your encounter: They know their own bodies, and know what they might need to be comfortable. Explain that most of the activities are designed to be done on the floor and find out if this would be difficult. Would it be easier to have all the participants in chairs for some of the activities? Will they need extra cushioning? Are they amenable to having someone help them get down on the floor if they need to lie down? If not, can they figure out a good workaround?

People on the autism spectrum may also struggle with receiving and giving the appropriate amount of touch. While people who are far to one end of the spectrum don't like to be touched (it overwhelms them), many of them do...they just don't know how to go about getting the touch they want. They will likely ask a lot of questions and may be reluctant to participate in some of the activities. The clear, direct communication we will be using will appeal because of the reliance on spoken language instance of body language and nonverbal cues. It can be a helpful tool for them to navigate social situations.

Nurturing human touch can be a godsend when recovering from trauma. While participants dealing with PTSD may not need any special accommodations during the activities, it's possible that they will need more support than other participants after your encounter. We will discuss this more in Chapter 14, Aftercare, but make sure they have professional support lined up in the event of a crisis.

You have carefully and thoughtfully made all of the arrangements to create a space that will be mentally and physically comfortable for your participants. And now, finally, we can begin to address your touch needs!

CHAPTER 12

Boundaries 101

USTIN'S DOWNTOWN IS sprouting skyscrapers faster than a cow pasture popping up mushrooms after a thunderstorm, and I drive by several large construction sites daily. For months, there will be a big hole in the ground that only seems to get deeper. I sometimes wonder if the building in the pretty picture is ever going to get built as I watch them add more rebar and concrete to the hole. And then, one day, a building frame rises from the hole, and within weeks, another luxury apartment building is born. The foundation is the most important part of the process.

This chapter about boundaries will be your foundation for building relationship structures of nurturing human touch. It will take less than an hour to construct, but just like a skyscraper, it is the most important part of the process.

This may seem tedious, ludicrous, and stupid, but remember, the goal here is to create a structured experience. Boundaries are simple, and yet most of us don't have a solid understanding of them. Once you start learning about creating clear boundaries, you will come to realize that (a) most people don't consider them, (b) most people have too few or too many, and (c) very few people can gracefully set or accept them.

A few years back, my friend Kelly was talking about their struggle to eat well in a world full of sugary, fatty, deep-fried foods that use bread as a delivery device. They likened it to living in 19th-century London and being one of the tiny percentage of people who realized that cholera could be contracted through contaminated water: while people were drinking water and getting sick, a few people stayed healthy by avoiding it. Avoiding it wasn't easy because this diseased water was everywhere, in everything.

Take this metaphor to heart: Once you learn this boundaries stuff (which, honestly, should be taught to kids in 7th grade), you will feel like you have to walk through a minefield of people who have the bad-boundary disease. On the positive side, you will learn to be more clear in your interpersonal relationships even when you're not touching anyone, and thus can avoid getting overly involved with people who wouldn't know a healthy boundary if it screamed in their face.

When we did the Snuggle Salons, Boundaries 101 was the thing people needed to know before we could turn them loose for some artisanal, free-range snuggling. People know how to hug or give a foot massage; what they don't know is how to ask for it, or say yes/no. I often heard repeat visitors say that what they learned during the boundaries part of the evening had spilled over into other parts of their lives, and they were happier for it.

While this part of the gathering may elicit some groans, or eyerolls, you need to do it. It will give you a framework in which to enjoy giving and receiving nurturing human touch. Doing it in a small group means it should only take 15 to 30 minutes to get the basics.

Opening Questions/Icebreakers

These three questions aren't really designed to teach you about boundaries, but they will get people thinking about touch. Optional: You can email them out prior to your first meeting so that people will begin to observe how people interact with each other in a tactile fashion.

Read the question, answer it yourself, and then ask each of the other participants to answer. Give people about a minute apiece to speak.

Question 1:

What are you hoping to get out of this encounter? Why do you feel like you need more touch in your life?

Question 2:

Do you think of yourself as a touch-averse, touch-sensitive, touch-situational, or touch-enthusiastic person? Can you think of a time when you encountered a person who needed more (or less) touch than you and it made for a difficult interaction?

Question 3:

Recall a time in your life when you gave or received platonic, nurturing human touch in a way that felt good for you.

Boundaries Exercises

This is where things get weird, awkward, and silly. I encourage you to get back in touch with your inner 12-year-old during these questions. Laugh nervously. Stammer. Cross your arms. Avoid eye contact. Turn bright red. Gulp loudly.

This stuff isn't easy the first time around. But hey, look on the bright side: All of the other people who are doing this are going to be just as embarrassed as you. And let's face it: You've probably seen each other in much more awkward states, like passed out drunk in the corner.

The goal here is to practice saying yes and no, asking for what you want, and negotiating. You don't need to put your hands on other people – for now it's just talking. I promise, it will get easier once you have done it a bit.

Exercise 1: SAYING NO

Ask the person next to you for something touch-related, and have that person decline.

Person 1: "Would you like me to touch your hair?"
Person 2: "No."
Person 1: "Thank you."

Person 2: "Would you please scratch my back?"
Person 3: "No."
Person 2: "Thank you."

Go around the circle until all of you have had a chance to both ask for something and say no to it.

Spend a few minutes asking people for their reactions and observations before moving on to the next exercise. Was it hard to hear no? Was it hard to say it? While being rejected is hard for many people, the feeling of rejecting another person can also be difficult.

The cool thing about this exercise? You can hear NO and you will still be alive. It won't be the end of the world, and it might not be the end of the relationship/encounter with the other person, especially if that person is your friend.

NOTE: If someone says no to you, always say thank you. You are thanking them for being honest about their needs. You are also thanking them for knowing their limits and taking care of themselves. By acknowledging that you respect their decision, you give them positive feedback and encourage them to say no when they need to in other situations.

Exercise 2: TUNING IN

Ask the person next to you for a type of touch you would like, or if they would like a particular type of touch. Once you ask, shut your piehole and wait patiently for the response. The person being asked will take a breath, maybe close their eyes, and take a moment to tune in to their gut to figure out if they really want what is being asked of them, or if they need to say no.

Example 1:

> **Person 1:** "Will you rub my feet?"
> **Person 2:** (thinks for a moment) "Yes."
> **Person 1:** "Thank you."

Example 2:

> **Person 1:** "Would you like a hug?"
> **Person 2:** (thinks for a moment) "No, not right now."
> **Person 1:** "Thank you."

Do this exercise 3 or 4 times so that all y'all can get a bit of practice with it. Many of you will have an automatic response: Maybe you were raised to be a people pleaser so you always say yes, or perhaps a bad encounter with someone who touched you non-consensually will leave you always saying no. Instead of blurting out your default response, take a deep breath and feel into your body. Do not feel rushed or pressured, and give others the same courtesy.

Exercise 3: NEGOTIATING

Next, we will learn how to reject an initial offer and counter with another offer in order to find something that is mutually enjoyable for both people.

Example 1:

Person 1: "Would you like to spoon with me?"
Person 2: "That feels a bit too intimate for me at the moment. How about if we hug standing up?"
Person 1: "I would love that, thank you!"

Example 2:

Person 1: "Would you like me to brush your hair?"
Person 2: "I just put a bunch of product in it and don't want to mess it up. Do you think you could scratch my back instead?"
Person 1: "Sure!"

Example 3:

Person 1: "Would you rub my feet?"
Person 2: "I'm weird about touching people's feet. Maybe we could hold hands instead?"
Person 1: "That reminds me too much of my ex-girlfriend – she loved to hold hands. How would you feel about stroking my face?"
Person 2: I could do that!"

Here we have gone from just figuring out how to respond to another person's request to thinking about what would really make us happy. The key to this is having many things in your nurturing-human-touch repertoire that would satisfy the need for touch in a way that feels good for both giver and receiver.

Generally, when one person is less comfortable with touch than the other one, it's a good idea to find a solution that will make the more uncomfortable person happy. Win-win is rare in our world; usually it's "my way or the highway." Getting to yes is a beautiful thing.

Exercise 4: CHOOSING AN OPTION

The question "What do you want?" can be daunting to answer: Many times we have no idea what it is that we want, especially when it comes to nurturing human touch. You can make it easier by giving options to choose from.

Example 1

Person 1: "Would you like a hug or your face stroked?"

Person 2: "Hmmmm....having my face stroked sounds really nice. Thank you!"

Example 2

Person 1: "Would you put your hands on my belly or run your hands through my hair?"

Person 2: "I would be happy to do both those things. Which would you like first?"

Example 3

Person 1: "I would like to scratch your back or give you a hug. Do either of these sounds good to you?"

Person 2: "At the moment, neither do. But thank you."

Person 1: "Thank you."

While it seems like there are only two choices, A or B, there are five choices: (1) option A, (2) option B, (3) options A and B, (4) neither option A nor B, or (5) something different entirely.

After you have gone through the exercises, ask people if they feel like they need more practice. Are they getting the hang of it? Do they have other observations or revelations to share?

When the discussion is finished, take a break. Your guests will be going through withdrawal pangs after not having communed with their constant and loyal companions – their phones – for a bit, and will want to check 'em. People may want water or the bathroom before going on to the physical touch portion of the day.

Are we ready to start making contact with each other? YES?!?! *Now* we can get to the hands-on activities!

CHAPTER 13

Sharing Touch with Your Friends

WHAT ACTIVITIES COME to mind when I say "nurturing human touch"?

If I asked a random assortment of people on the street to name five different sexual activities, they could rattle them off without even thinking about it. I might get some dirty looks, and some folks would mumble "pervert" under their breath, but it would be easy for people to do.

If I asked these same people to name five different ways that you could give someone nurturing human touch, their list would stop at "a hug." The only people who might do a bit better on this survey would be parents: Touch is one of the ways they care for their children. Despite this, they likely wouldn't think about touching their friends like they would their offspring.

And that's a pity.

At this point, you probably can't think of five ways you could give nurturing human touch, but we're about to change that. I've put together a group of activities for you to work through that are going to give you more than five options. By the time we're done, you will have many things to offer and receive that will scratch that itch for kind, tender physical contact.

The sequences you are going to go through are designed to do five things:

* Give you a wide assortment of ways you can make physical contact with other humans.

* Get you into the habit of asking for consent before you touch somebody, and giving consent before somebody touches you.

* Allow you to acclimate yourself to touch slowly before trying more intimate forms of touch.

* Offer you an opportunity to experiment with giving and receiving touch, and giving and receiving feedback about touch.

* Provide you many chances to laugh, play, and be silly. Sometimes they will give you a chance to be sad, cry, and be compassionate.

A couple of things you need to know before going through these activities:

* Many of the activities are designed for two people, but they should still be done in a group setting with people paired off. Remember that the goals here are to build a common framework for giving and receiving touch, and to reframe it as a social activity instead of a sexual activity. After we go through puberty, the idea of lying down with another person with bodies entwined conjures up images of sex. It will take a bit of time to work through that programming and return to a place of playfulness and innocence. Unless you have had a lot of public/group sex, this will be easier to do if there are more than two of you in the room. We want to keep the ambiguity and assumptions to a minimum.

* It's important to go through these activities in order. You may be thinking to yourself, "Touching someone's hand is stupid – I really need a hug!" Remember, though, that one of your companions might be thinking, "Touching some-

one's hand...NOPE." In my experience, the people in the latter group start experiencing nurturing human touch and realize that it feels good. After that light bulb goes off in their heads, they may be open to trying some of the activities involving more intimate touch. Or not.

✳ It's entirely possible that some people won't feel comfortable with a particular activity. *That's perfectly fine.* Do not pressure or shame your friends; let them set their own boundaries. If it's an activity that needs to be done in pairs and one person decides to sit it out, one of the people from another can fill in and give and receive twice.

✳ Activities are grouped into three sets or, as the hipsters say, small batches. You can choose to go through all of them in one day, or go through the first set and take a break. If you go through all three sets plus the Boundaries 101 stuff, you should set aside four hours.

✳ At the end of each activity, I have listed some variations. While you can try the variations during the activities, their main goal is to give you ways to modify the activities to be done in different settings, at different times. For sake of brevity while you are learning, you probably want to stick to the one activity given, and move on to the next one.

You are now armed with information about touch, boundaries, and consent, and the only thing left to do is start giving and receiving nurturing human touch! Shall we begin?

First Set

1-1: HAND CARESS
# of People: 2
Giver/Receiver: Yes
Length of Time: 6 minutes

Designate one person as giver, and one as receiver (you will switch off for the second part). Set the timer for two minutes. The giver will ask the receiver, "Would you like me to touch your hand?" If the answer is yes, the giver will take one of the receiver's hands in both of their hands, close their eyes, and then slowly begin lightly touching the receiver's hand. This can be done with the fingertips, or a full, open palm. The receiver should close their eyes as well.

The giver focuses on learning what it feels like to touch lightly or firmly, quickly or slowly. Pay attention to the various textures of the palm, the back of the hand, the fingernails and the fingers. There is no specific goal here other than having a bit of human contact, and getting a feel (no pun intended) for pressure and timing.

The receiver will focus on what it feels like to be touched. Is it calming? Alarming? Does it tickle? Feel good? Is it itchy? Sharp? How does the rest of the body respond to this small amount of touch? Did they forget to breathe?

After the two minutes are up, each person should chat about their experience. Set the timer again for two minutes, and switch the giver and the receiver. Lather, rinse, repeat.

Variations on the hand caress:
* Holding hands across a table
* Holding hands while walking,

* Holding hands while sitting
* Holding hands while standing
* Shaking hands, and then taking the person's hand in both your hands when finished

1-2: FOOT MASSAGE
# of People: 2
Giver/Receiver: Yes
Length of Time: 12 minutes

Again have a giver and receiver. Set the timer for two minutes. The giver will ask the receiver, "Would you like me to rub your feet?" If the receiver answers yes, the giver will take one of the receiver's feet in both of their hands to start, close their eyes and then slowly begin lightly touching the receiver's foot. Because feet can be ticklish, a deeper touch that resembles a massage might be wanted – ask the receiver if they would like you to apply more pressure. You can touch the toes, the soles and tops of the foot, the heel, the Achilles tendon, and the ankle. Having a bit of lotion or cream can be helpful. (Super pro tip: Do *not* be like me and say, "It rubs the lotion on its skin" in your best Buffalo Bill voice while you do this...unless your friend would appreciate it....)

After the two minutes are up, reset the timer and do the other foot. When both feet have been done, each person will talk about their experience. The cycle then gets repeated with the receiver rubbing the giver's feet.

Variations on the foot massage:
* Two people each rubbing one of a third person's feet simultaneously
* Pedicure

* Holding someone's feet while it's cold, to warm them up
* Foot washing (Karuna Sessions tested, Jesus approved)
* Putting on someone's shoes and socks for them

After this activity, give people the opportunity to wash their hands.

1-3: TOUCHING HAIR/EARS/HEAD
# of People: 2
Giver/Receiver: Yes
Length of Time: 6 minutes

Again have a giver and receiver. The receiver can sit on the floor or sit on a chair (whichever feels most comfortable). The giver can sit to one side or the other of the receiver, or stand behind them. Set the timer for two minutes. The giver will ask the receiver, "Would you like me to touch your hair?" If the answer is yes, the receiver will close their eyes. The giver should probably keep their eyes open for this since they will need to be able to see where they are putting their hands!

The giver will begin to run their fingers through the receiver's hair and rub their scalp with their fingertips. They can pull the hair back from the receiver's face and run the fingers along the ears, or pull the hair up from the nape of the neck and place a hand there. If the receiver is bald, gently touch their scalp, ears and back of the neck.

The giver should get feedback: Ask the receiver if they like what they are experiencing, and if the receiver would like anything different.

After the two minutes are up, each person will talk about their experience. The cycle then gets repeated with the receiver touching the giver's hair, scalp, ears, and nape of the neck.

Variations on playing with hair:

* Brushing the hair
* Braiding the hair
* Dying the hair
* Shaving the head

1-4: CARESSING THE FACE
# of People: 2
Giver/Receiver: Yes
Length of Time: 6 minutes

Again have a giver and receiver. The receiver should lie on the floor or sit on a chair (whichever feels most comfortable). The giver can sit next to the receiver or stand to one side or the other of them, facing them. Set the timer for two minutes. The giver will ask the receiver, "Would you like me to touch your face?" If the receiver says yes, they will close their eyes.

The giver will begin by placing an open hand on one of the receiver's cheeks, and then begin gently touching their receiver's face. They can use their fingertips (or a single fingertip), open hand, or the back of the fingers to stroke the receiver's forehead, cheeks, chin, brows, lips, and nose. Stopping to rest the hand in one spot is nice as well. If it helps, the giver can visualize themselves as a parent and the receiver as a child who is being tucked into bed.

After the two minutes are up, each person will talk about their experience. The cycle then gets repeated with the receiver touching the giver's face.

Variations on caressing the face:

* Applying a face mask
* Applying lotion or cream
* Shaving the beard

These four activities comprise the first set. This is a perfect time for people to take a break, use the bathroom, drink some water, etc. You may also choose to end your encounter here, and take up the next couple sets of activities on another day so that people have some time to process their experience, and see how it makes them move differently through the world.

Second Set

If you are coming back to the second set of activities on a different day, take time to do a brief check-in and find out how people are doing. Did they enjoy the experience the first time around? Did they notice a difference in their mood, energy levels, or interactions with the world after their first encounter with touch? Or did they feel foolish, ashamed, or embarrassed? Did they talk to other people about what they had done, or was it something that they didn't feel like they could mention?

After the check-in, ask if people want to go over any of the boundaries exercises again. If not, begin....

2-1: I'VE GOT YOUR BACK
# of People: 4
Giver/Receiver: 3 givers, 1 receiver
Length of Time: 5-8 minutes

The receiver can sit on the floor or in a chair. The receiver should then recount an experience from their life that was hurtful (an offhand comment from someone on the street, something a boss said, an argument with a friend, something recent, or something that has stuck with them for years). The receiver will take no more than a minute to talk about their experience. The other members of the group will simply listen.

Next, one member of the group will either sit with their back against the receiver's back, or stand behind the receiver and put their hands on the receiver's shoulders. The receiver should talk about another experience they had in which they felt unsupported or hurt.

After the receiver has done this both ways, the giver should become the receiver. After the receiver tells their story alone, one of the other group members will then sit back to back with them or put their hands on their shoulders. Go through until each member of the group has had an opportunity to talk about a problem with and without receiving physical touch support.

After everyone has had a turn, discuss the experience as a group. Did it help to have physical contact while talking about problems? Did you feel supported, or not? Was it easier or harder? Did it feel more or less vulnerable?

2-2: CIRCLE OF SUPPORT
# of People: 4
Giver/Receiver: 3 givers, 1 receiver
Length of Time: 10-20 minutes

The receiver will lie on the floor or a massage table on their back. They will then direct the three givers to where they want each giver to be: at their feet, head, sides, etc. The receiver will next ask each giver for a specific type of touch in a specific area. ("Will you place your hand on my belly?" "Can you put your hands above my face without touching me?") Set the timer for two minutes. The givers may experiment with various types of touch, and ask for feedback from the receiver. The receiver may ask for different types of touch, or ask the givers to keep their hands still. When the two minutes are up, another person will be the receiver. Repeat the process until each person has had an opportunity to be in the center of the circle.

If you have time, you can go through the cycle again with the receiver lying on their stomach. The back body and the front body will respond differently to different types of touch, and it's fun to experiment with what that feels like.

When each person has had a chance to be the receiver, have a debrief. Was it comforting or disconcerting to be touched by so many people at once? Was it easier to give than receive? Were you able to figure out what you wanted? What would you do differently next time?

2-3: RELAX AND REFRESH
# of People: 2
Giver/Receiver: Yes
Length of Time: 25 minutes

The receiver will lie on the floor or a massage table on their back. The giver will ask "Would you like me to use light touch to relax you?" If the answer is yes, the timer starts for 10 minutes. The giver will run their fingers or hands lightly over the legs, arms, torso, feet, hands, and head of the receiver. The giver might also rest their hands on various points and be still. At some point, the giver should switch to the other side of the receiver to more easily access the other side of the body. The idea is to lightly touch as much of their body as possible (though still avoiding the erogenous zones, of course). The giver also has an opportunity to experiment with different types of touch, and ask the receiver for feedback.

When the 10 minutes is up, the giver and receiver switch places, and the cycle repeats.

When each person has been the receiver, discuss the experience. Was this comforting or did it make you nervous? What happened to your breathing? Could you feel a difference between the two sides of your body if one side received touch before the other side?

2-4: TOUCH FOR INFLAMMATION OR INJURY
# of People: 2
Giver/Receiver: Yes
Length of Time: 12 minutes

The receiver will tell the giver where they are in pain. (It could be a stiff neck, a sprain, inflamed knees, sore lower back, tired feet, headache, etc.) The receiver will then position themselves (sitting, or lying on back or stomach) in a way that allows the giver to touch the area that is in pain easily. (In this example, we're using the knees.) The receiver will then ask, "Will you touch my knees?" If the giver says yes, the timer starts.

The giver will spend five minutes gently and lightly touching the area with the tips of the fingers, covering both the area that is in pain (knees) and the shin and the thigh within about 6" of the knee. The giver can also surround the tender knee with their hands at different points. Do not use deep touch or massage.

When the five minutes are up, the giver and receiver switch places.

After each person has received, discuss the experience. Did it feel good to have someone give attention to your physical aches and pains? Did you have better range of motion when it was done? If there was swelling or inflammation, did it lessen?

Congratulations! You've finished the second set of activities. It's now time to take a break or call it a day.

Third Set

If you are coming back to this practice on a different day, do a check-in, and make sure people are still in agreement around boundaries.

This set of activities will involve a great deal more physical closeness, and crosses into ways of being affectionate that are reserved for people in our culture who are romantically involved with each other. Make sure that everyone knows that this will be happening and that if they need to opt out, they should feel comfortable doing so.

3-1: HUGGING WHILE STANDING UP
Number of People: 2
Giver/Receiver: No
Length of Time: 5 minutes

Pair off and stand facing your partner. One person will ask, "Would you like a hug?" If the answer is yes, hug each other as you would if you were greeting a friend (probably 3-5 seconds). Allow the person who answered "yes" to pull away.

Stand facing your partner again. The other person should now ask, "Would you like a hug?" If the answer is yes, set the timer for 30 seconds and give each other a 30-second, standing hug. Separate when the timer goes off.

Stand facing each other for a third time. The first person will again ask, "Would you like a hug?" When they get a yes, set the timer for 60 seconds and hug for a full minute. When the timer goes off, separate.

Discuss your experience. Did your partner relax, or tense up? Was it hard to sustain a hug for an entire minute? Were you desperately waiting for the timer to go off, or did you feel like you could get lost in it?

3-2: THE HUMAN CHAIR
Number of People: 2
Giver/Receiver: Yes
Length of Time: 12 minutes

The giver will sit on the floor with legs spread in a V. If necessary, the giver can have their back to the wall for better support. The receiver will sit in front of the giver, with their back to the giver, legs out or crossed, between the giver's legs. The giver will ask, "Would you like to lean back on me?" If the answer is yes, start the timer for five minutes. The receiver can place the giver's arms where they want them – around their waist or shoulders. The giver may clasp their hands around the receiver, or they can ask the receiver if they would like their belly or arms touched. The giver can also stroke the receiver's hair or face. When the five minutes are up, switch positions and do it all over again.

When you have finished, discuss the experience. When you were giving, did you avoid adjusting your body out of an uncomfortable position so as to not upset the receiver? Did you feel safe sitting like this, or nervous? Did you find it relaxing or difficult?

3-3: SNUGGLE SANDWICH
Number of People: 4
Giver/Receiver: No
Length of Time: 25 minutes

Set the timer for five minutes. One person in the group will lie on their side on the ground. The next person will lie down right behind them and place their bottom arm under the first person's neck or waist or upper arm, and the top arm over the person's waist. Continue until all participants are lying back to front. Start the timer. Participants may talk about the experience or something else altogether, laugh, sigh, cry, or relax into silence. If someone starts to get fidgety, ask them if they need their arm positions adjusted. When the five minutes are up, the last person to lie down will move to the front of the line. Let the timer go for another five minutes, and switch around two more times. Each person in the group will have the experience of being in the middle twice and on the ends twice.

When you have finished, talk about your experience. Did it feel comforting or unsafe? What was the difference between being in the middle and on the end? Was it relaxing to be surrounded by bodies?

3-4: SPOONING
Number of People: 2
Giver/Receiver: No
Length of Time: 12 minutes

Choose which person will be the big spoon and which one will be the little spoon. The person who is the little spoon will lie on their side on the floor. The person who is the big spoon will ask, "Would you like me to spoon you?" If the answer is yes, the big spoon will lie on their side right behind the little spoon (as they did in 3-3). The little spoon will let the big spoon know where to place their lower arm – under their neck or waist. Start the timer for five minutes. When the time is up, switch places and repeat for five minutes.

After you have finished, discuss the experience. Did it feel safe, or too sexual? Did it feel strange to be the big spoon if you were a smaller person than the person you were spooning? If the person spooning you was smaller than you, did it feel comforting? Which position did you prefer, big spoon or little spoon?

3-5: FACE-TO-FACE HUG LYING DOWN
Number of People: 2
Giver/Receiver: No
Length of Time: 6 minutes

Two participants lie on the ground, facing each other. One participant should be slightly lower than the other so that their head can rest on the chest/shoulder of the other person. Set the timer for five minutes. Participants may stroke each other's faces or shoulders or back, or they might simply lie still.

When the time is up, discuss the experience. Did this feel nurturing or erotic? Did you find yourself having romantic feelings, or did you feel like you were keeping the other person safe? Was it easy to relax into the experience or did you need to stay guarded? Was it hard to be present?

Congratulations: You've just graduated from Miss Epiphany's Finishing School for Nurturing Human Touch! Please exit through the gift shop and peruse our tchotchkes on your way out.

After going through your up-close-and-personal encounter with nurturing human touch, you will probably feel great. It's possible, though, that you have opened a Pandora's box of overwhelming emotions in yourself, or others. In the next chapter, we'll take a look at Aftercare.

CHAPTER 14

Aftercare

NURTURING HUMAN TOUCH is fantastic for your body, mind, heart, and spirit. It is simple, yet profound...and it is a great responsibility. As the leader of this expedition into uncharted territory, it will be up to you to look out for your fellow travelers' well-being when y'all return to the default world.

Sharing touch after going without for a long time can make you drop your guard. All that oxytocin makes you feel open and loving. You may find yourself talking to people in public where you would normally keep to yourself, or smiling at everyone in your office. But after a couple of days, the warm feelings may fade away, and you find yourself wondering if it was all a dream. You will also be trying to figure out how to get back to where you were.

This feeling will be familiar to anyone who has had a connective, immersive experience at a festival, show, workshop, conference, or event. Sharing something unique with like-minded people is powerful, especially if the behavior and actions are at odds with our default day-to-day interactions.

We discussed ways you can make people feel safe in Chapter 9, and how to set up your space in Chapter 11. These things will help you and your participants have a good experience, and will help make a graceful transition back to the real world. Below you will find some more suggestions on how to take care of your fellow touch explorers after you've finished your expedition.

Use Your Words

When you have finished up with your last activity, have everyone go around the circle and talk about their thoughts and feelings. If they need some prompting, ask them to give you one word to describe their experience, and then go from there.

Let them know how much you appreciate them for showing up, taking a chance, and trying something unorthodox. Tell them that they should call you if they need someone to talk to, or if they are struggling with adjusting to the default world where people avoid touching each other.

Even if everyone is happy and satisfied and ready to do it all over again, calling them to check in with them the following day is kind and thoughtful. They may feel isolated and lonely. Talking to someone who has been through the same experience, and understands why they are feeling a bit blue, can help rekindle the warmth and connection. It will be good for them...and for you.

If you hate the telephone, you can use text messaging, but they will appreciate hearing your voice. Video chatting is nice too, so they can hear your voice *and* see your face.

Break Bread

If you go through all three sets of activities in one day, chances are you will have worked up an appetite. Sharing a meal is a perfect way to transition back to your normal life and celebrate your newfound connection. Don't forget to drink plenty of water.

Getting together for lunch or dinner a few days after your experience may be nice, either one-on-one or as a group. Extra-added bonus: you will have another opportunity to share some nurturing human touch.

Self-Care

Before parting, remind your participants to pay attention to what their body or heart needs over the next couple days. A bath, a walk, a nap, a movie, or a conversation with friends are simple forms of self-care. Most importantly, don't forget to schedule in your own self-care. You just did a lot for others by holding the space and organizing the encounter, and you may be feeling worn out.

Plan Another Encounter

If you enjoyed getting together as a group, you may want to find a date for another encounter immediately. It can be once a week or once a month, but letting your friends know that you're not treating the experience like a one-night stand will help them feel cared for and seen. Planning ahead is also helpful because most people are very busy, and getting things on the calendar far in advance makes it easier to commit.

When Things Go Awry

If there is a potential for this experience to open a floodgate of emotions with your participants, they will need to have support lined up to process it. Most of us are not trained therapists, and you don't want to find yourself in a position of dealing with a crisis you are not equipped to handle. It is not always possible to predict how an experience will hit someone, but if you know that someone in your group has a history of trauma, you will want to make sure they have a plan.

Set and Setting

The phrase "set and setting" was popularized by Timothy Leary in the 60s. This concept is used to ensure that a psychedelic trip goes well. It's in everyone's best interest that the tripper has a positive experience. We talked about "setting" at length in Chapter 10, but "set" is equally important.

"Set" refers to the tripper's mental state and determines whether they have a heavenly or hellish experience. Before would-be psychonauts swallow that pill, they ask themselves if they are in the right frame of mind to have their perception altered. Have they recently experienced trauma or turmoil that might send them down dark corridors of their mind? Will they get caught in a whirlpool of guilt and regret that will make them wrestle with their demons? If they are not in the proper headspace, it's entirely possible they will end up in the med tent getting shot up with thorazine. Not. Fun. (Trust me. It sucks.)

Pay Attention to "Set"

While you likely have a good idea of what your close friends have been through, and what their struggles are, you may not know everything. Prior to getting together, ask them if they have any known traumas or mental health issues that might be triggered by their experience of being touched.

If they have something ongoing, find out if they are seeing a therapist, and if they have talked to said therapist about their lack of nurturing human touch. Planning your expedition into nurturing human touch a day or two prior to their therapy appointment can be helpful, or ask your friend to schedule a therapy session after your get-together.

If your participant knows that someone will catch them if they fall (so to speak), that may be enough to keep a breakdown from happening.

Part 2 of this book has provided you with step-by-step instructions on how to share touch with your friends. But how will this translate to the real world? Part 3 will focus on taking what we've learned into different experiences, situations, and phases of your life.

Part III

The Real World

Hunter

Last spring, my friend Hunter stopped by my house. My boyfriend had recently moved out and Hunter and I were enjoying some one-on-one time. Unfortunately, my boyfriend had left behind his big, sagging, beige leather sectional and we sunk into it uncomfortably. We shifted from side to side, trying to find a way to keep it from hurting our backs while we chatted.

We discussed Hunter's plans to return to the west coast. They had recently found a copperhead on the steps of their house, and woken up to a bed full of fire ants a few days later. They were feeling like Texas was a hostile place, and they wanted to be in an area with fewer dangerous creatures.

I had my feet up on the edge of the table, and Hunter commented on my bright red toenails. I told them I had just gotten a pedicure.

"A pedicure? That's where they rub your feet and stuff," they said. Not only is Hunter not much for pampering, but they were down to selling plasma for grocery money. A pedicure was not in their past, and probably not in their future.

"Here, take off your shoes and I will rub your feet for you," I offered. Hunter laughed a bit nervously, but nodded affirmatively. They took off their worn work boots and put their feet in my lap.

As I began to rub Hunter's feet, they lay back on the arm of the couch, closed their eyes and sighed. "That's nice," they said. I know quite a bit of Hunter's history, and it includes violence, abuse, and betrayal. They have many compelling reasons to distrust people, and their primary relationship was with their dog. They don't get touched by another person very often.

"You are pretty much the only person who I let touch me," Hunter said shyly. "Most of the time I even avoid hugging peo-

ple, but when I do, I'm reminded that I actually do need touch in my life."

I continued rubbing their feet for another ten minutes, and then held each foot in both of my hands. Hunter lay quietly, with their eyes closed. I gently lifted their legs, moved out from under them, and lowered their legs back down onto the couch. I stood up, looked down at Hunter, and brushed their hair back from their eyes.

"You rest for a bit, sweetie," I said softly. "I'll grab us some water."

"Thank you," they murmured, and smiled.

CHAPTER 15

What's Next?

IF YOU'VE GOTTEN through this process, congratulations! Even if you are like "What the fuck was that? It's way too weird, and I'll never do it again," you tried something new, discovered that you don't need a lot of touch, and learned more about boundaries.

Or maybe you will decide, "I'm going to find my soulmate and they will give me all the touch I need." That's totally your prerogative. I hope you meet the person of your dreams; when you do, you will have some new things to try with them and will be able to ask for what you want.

But if you and your friends are delighted with the results of this social experiment, there are many ways you can integrate touch into your relationships going forward.

* Commit to meeting your nurturing human touch needs on the regular. Set aside some time once a week/month to get together and continue to share touch as a group. Each person can say what they want or need in the beginning and set up accordingly, or you can start out with a snuggle sandwich and get more specific from there.
* Share touch one-on-one with one of your fellow explorers in a public or private setting: cuddling on the couch to watch a movie, or hanging out on a blanket in the park.

✳ Plan an evening where you take turns focusing on each person through touch, or you care for one person who is going through a difficult time.

✳ Have an evening of social grooming with face masks, hair care, and nail painting.

✳ Ask one of your crew to attend an appointment with you. Have them give you touch support while you are receiving the results of a medical test, or getting other difficult news.

✳ Ask one of your crew to hold you or rub your feet when you are talking about your problems or trying to process something. You can also offer the same.

While you can be somewhat less formal, consent, asking for what you want, and negotiating are going to be de rigueur in any situation involving touch. Remember that you can ask to receive *and* give. This is an important distinction, and you will get different responses, as you move forward.

Asking to receive touch from someone who doesn't have the same frame of reference about touch requires us to take a slightly different approach. Up until now, we have talked about consent in terms of making sure that the other person wants and needs touch. When we are talking about getting our own touch needs met, we will be be focusing on our own needs first. The consent part comes in a bit later, when the other person has the opportunity to say yes or no to giving us what we are wanting.

Vulnerability

Vulnerability has been a negative word for centuries. Its Latin root, *vulnus*, means to wound, either physically or emotionally. In the past few years vulnerability has gained traction as a strength, and something that is seen as a desirable quality in romance, friendship, management, and parenting. The ability to open up and reveal the things that we fear and avoid has become a must-have life skill.

Brene Brown has written at length about vulnerability. "Staying vulnerable is a risk we have to take if we want to experience connection," she says in her book *The Gifts of Imperfection*. Shame often keeps us from asking for what we want or need. Getting more nurturing human touch in your life will require vulnerability, and taking risks.

Starting with vulnerability in your ask is the emotional equivalent of a dog rolling over and showing you its belly. When you let down your guard and move past the social pleasantries of "I'm doing great today, thanks for asking!" you are telling the other person that you trust them not to harm, manipulate, or attack you with the information you are revealing to them.

A vulnerable ask for touch could look something like this:

"I have had a really rough day and am feeling super stressed. May I have a hug?"

"I just got some bad news about my brother, and could use some comfort. Would you be willing to hold my hand for a moment?"

"Last week I hit reply-all on a sensitive email and my boss asked me to meet with him in his office tomorrow morning. I am freaking the fuck out, and could really use a hug, if you have one to spare."

There is no magic word combination or special body language that will automatically get you what you want: Every situation and every ask will be different. So many different factors to contemplate: your relationship to the person you are approaching, social vs. work setting, if you're hungry or tired, or dozens of other variables. Be honest, be willing to take no for an answer, and if somebody does say no, don't take it personally.

Remember: It's always okay to ask. And it's always okay to say no.

Sharing Touch with the Non-Initiated

If you are the type of person who likes to proselytize, you may find yourself extolling the benefits of nurturing human touch. Everyone who is within earshot of you will hear about how awesome it was when you and your friends got into a snuggle sandwich and giggled like little kids while watching *Game of Thrones*. You will run around with your arms spread like a love zombie, but instead of saying "Braaaaaaiiinns" you'll be saying "Huuuuugggsssss."

And people will start slowly inching away, and begin avoiding you like you're trying to indoctrinate them into your cult. Have you had a friend who discovered polyamory and talked about nothing else for months? Yeah, you know the drill. Don't be *that* person.

Remember that the vast majority of people are swimming in the touch = sex fishbowl, and you have escaped like Nemo to swim in the ocean of nurturing human touch. You have a new set of skills, awareness, and knowledge that other people may not understand, but the rules apply, even more so. *As the person who has a greater understanding of touch and boundaries, you have a greater responsibility to be more ethical and cautious with others.*

If you want to offer someone nurturing human touch, you don't have to be like a Jedi and use mind control; simply asking if they would like a hug will suffice. Make sure you give them plenty of time to respond. Remember when you were just starting out, and you had no idea what you wanted? Put yourself in their shoes for a moment! Choice is always better than force. And because you've had practice with rejection, you will be able to take "no" for an answer with grace and kindness. Being respectful of other people's boundaries is a tiny way to make the world kinder and more trusting, and to help change the cultural narrative that touch is bad.

Most of your interactions with the uninitiated will be of the hug variety. Hugs allow you to be both giver and receiver, can be done almost anywhere, and don't require much time. Foot rubs are also a great starting point; because you will be *all* the way at the other end of their body, they are one of the less intimate forms of touch...unless the person whose feet you're rubbing is Marsellus Wallace's wife in *Pulp Fiction*. That might get you thrown over a balcony into a greenhouse.

You may not have too much trouble finding people who do crave touch; touch hunger is rampant, and people are starting to recognize the importance of touch for health and happiness. You may also find that people respond to you differently: When you are getting your touch needs met, your body will be more relaxed and you will come across as less desperate. You will be laughing and smiling and happy, and people will wonder what drugs you're on (oxytocin > oxycontin). I suspect you'll find people who want to know more, and are happy to jump on your bandwagon without much coaxing.

One-on-One

You may go through the activities in a group setting, and then you discover that you have one other friend who wants to learn. Or you've approached several people about trying this, and all of them have looked at you like you've grown a third head...except for that one person who hasn't been on a date in six years and they're eyeing you like a puppy thinking you have a treat in your hand. Should you share touch with them?

That is entirely up to you. The boundaries/negotiation practice questions are doable one-on-one, but they aren't nearly as much fun. And you could certainly run through the first set of activities in a very public spot, like a park, so that it doesn't feel like something romantic. And you can always ask them to read this book so they understand what you're talking about.

It may be a gamble worth taking. Once a person begins to feed their touch hunger, they will calm down some. And if you model good, clear boundaries around negotiation and consent and asking for what you want, they will follow your lead. You will have teaching and learning opportunities aplenty.

Another option may be to go through the boundaries and hands-on exercises once every three to six months. The more people who join your nurturing human touch pod, the more options you will have to get your touch needs met!

Between Romantic Partners

As you may have already discovered, having sex with somebody does not mean that you will have as much touch as you would like. If you sleep wrapped up in each other's arms every night, you can have your touch needs met while you slumber, and most sexual activities involve a lot of skin-on-skin time. But these things are not a given in a relationship.

Whether you are in those first three months where you can't get enough of the other person, or you've been married for 40+ years, your relationship will benefit by consciously sharing nurturing human touch, and by delineating it as something that is distinct from sex. It can help you retain feelings of closeness and intimacy when you're fighting or when you need support in your personal challenges. It is super-duper important when one of you is sick/injured/busy and sex is off the table.

If you're interested in having more nurturing human touch in your romantic relationship, plan to have a conversation with your partner about your needs and expectations instead of leaving it to assumptions. Put a hugging ritual in place: Always hug each other before you walk out the door or when you come in at night. When you've had a rough day at work, ask for a foot rub or some snuggling while you talk about it. And if you want to have sex but you're both exhausted or stressed, the 10 Minutes of Relaxation activity in Chapter 12 is a fantastic way to connect...and it may give you the energy to knock boots.

Outside of Romantic Relationships

While people in poly relationships will roll their eyes and say "duh" at this suggestion, monogamous folks may be shocked. Don't write it off so quickly. There are many times in romantic relationships when receiving nurturing human touch from someone other than a spouse or partner can benefit the recipient and ease pressure in the relationship.

If you've been in any sort of long-term relationship, you already know that your partner may not always be able to give you what you need. It could be lack of bandwidth, or different desires for touch. Figuring out how to get touch needs met without resorting to cheating could be a godsend for millions of unhappily married couples. You can also participate in it in a social/group setting together to make it more understandable for the person who has a lower touch drive.

Going this route still requires having open, honest conversations with your partner, and oftentimes it does not go over well with the partner who has less desire for touch, or who has withdrawn from the relationship. All the more reason to identify your own desires and talk about them early on! Many couples have found *The Five Love Languages* framework to be a great jumping-off point to talking about what sorts of gestures make them feel loved and cherished. It can save you tons of grief as you try to navigate the curve balls of life while keeping your intimacy alive.

In Support Group Settings

While psychotherapists are barred from touching their clients, clients may be able to touch each other as peers. Support groups and 12-step meetings are a perfect forum to give people extra backup and care via nurturing human touch while they are struggling.

Before you try this, check with your group leader (if there is one) to make sure there is no policy against it. Avoiding romantic entanglements is often a suggestion in recovery, and it's possible that touch is verboten in your particular circle. If not, and you have gotten the go-ahead from your group leader, you can start by asking someone to stand behind you with their hands on your shoulders, or to hold your hand, while you speak. Others may or may not follow your lead, but if you find it helpful when you are sharing something painful, it is worth the effect.

As a Part of Ritual

Oftentimes, the goal of ritual is for the greater good, marking the passage of time or creating energy and releasing it to the universe. If the focus of a ritual is letting go of or calling in things for the individuals involved, touch is a tangible way to support the participants.

You can use touch to cleanse or rid the other person of energy or psychic cords by using your hands to brush and pull "gunk" off their body. Washing a person's hands or feet is also a literal and symbolic way to cleanse them. When it comes time to call in something fresh or new, one person can rub cream onto the other person's hands as they speak. And having all participants put their hands on one person while speaking their wishes and blessings for the person in the center adds an extra dimension of solidity to the work.

For Caregivers

While many couples could benefit from getting their nurturing human touch needs met by someone other than their romantic partner, people who care for others could use the support more than most. Many available jobs involve service or caregiving for low pay, and these professions have a high rate of burnout.

When we started Karuna Sessions, I was surprised to discover that one of our most enthusiastic groups of clients was new mothers. They absolutely loved having us do for them what they spent all day doing for their own children, and equated it to having their nurturing bank accounts refilled after months of incessant withdrawals

Many mothers talk about being "touched out" when they spend their days being pawed by grubby little hands attached to demanding children. After they received touch that nurtured *them*, they returned to their familial responsibilities feeling refreshed, relaxed, and present. As many of us raise our children in isolating nuclear families, away from extended families that may share in caregiving, nurturing human touch can go a long way toward making mothers feel supported.

Many of you will also go through the experience of being a caregiver when a spouse or parent falls ill. This can be especially difficult when you are taking care of someone who was a source of emotional, physical, and financial support: Suddenly all the responsibility and decision-making defaults to you. Getting away for an hour or two to have someone touch and hold you when you're overwhelmed from keeping track of everything lets you have a concentrated experience of self-care that will sustain you long after the actual physical touch has ceased.

For the Elderly

The older we get, the fewer touch opportunities we have, and the more we need it.

A generation or two back, our elders lived under the same roof as their children and grandchildren, and had opportunities to receive touch through childcare. Those days are long gone for many people. Older people are likely to suffer from the diseases that are exacerbated by loneliness. Many of the health benefits of touch and oxytocin we discussed in Chapter 4 can help with the physical and emotional pains of aging.

My friend Mickey is in their early 70s. A few months back, we were at a party and I was talking to them about my work. They asked me show them, and I began to touch their face. As I gently touched their forehead and cheekbones, their wrinkles smoothed out, they smiled and began to relax. It was easy to see the child they had once been. They enjoyed it so much that they asked me to touch their face several more times when our paths crossed that evening.

When you are touching older people, be cognizant of their physical limitations: Touch to the face, head, arms and hands may be easier for them than receiving a hug. Human skin gets thinner as we get older: strong pressure is not advised. IMPORT-ANT: Older people often have compromised immune systems. Make sure you wash your hands thoroughly prior to touching them, and avoid contact if you are sick and contagious.

During Illness

If aliens were to intercept our media feeds, they would be convinced that our planet was inhabited by young, healthy people. We have an awkward relationship with illness and disease. While vaccinations, antibiotics, and fresh food and water have vastly reduced maladies that cause sudden death, they have been replaced by illnesses with long declines (like cancer or Alzheimer's) that are painful for both the sick person and those who love them.

Being sick can be a physically isolating experience that includes lots of unwanted and unwelcome touch from strangers. Sequestering people in hospitals or care facilities leaves them alone and away from the communities and connections that may sustain them. This makes nurturing human touch vital and necessary during these times.

If you are recuperating from an illness, asking for nurturing human touch from your visitors will provide you with emotional care and support. With oxytocin's anti-inflammatory and immune-boosting properties, it may also speed up the healing process on a physical level. The first set of activities in Chapter 12 can be gently shared in a nursing home or hospital. If your immune system is compromised and touch would harm you more than help you, refrain from asking. Likewise, if a disease is contagious, you will want to protect your friends and avoid physical contact.

Death and Grieving

If our culture has a poor relationship with sickness, our relationship with death is downright dysfunctional. In the recent past, our relatives, young and old, would have died at home, and death was a familiar part of life. As health and life expectancy has improved, death has become something that happens behind closed doors and out of our sight.

If your relative dies in the hospital, a quick phone call to a mortuary can allow you to have their body "disposed of" without ever confronting the reality of their permanent absence. Because we outsource the care of dead bodies, and keep death at a distance, you may find yourself blindsided when somebody close to you dies.

Having an opportunity to touch and say goodbye to the physical body of someone you love, during and right after they die, can help you process your grief and begin finding closure. Nurturing human touch can also help you create deeply intimate memories of the person who is dying. I am thankful that I was home the weekend my father died, and that I was able to hold his hand, tell him that I loved him and that it was okay for him to go. I will never forget how small and frail he was after years of dementia, the sounds of his labored breathing and the feel of his delicate skin under my fingers. The experience would have been much more painful if I had arrived after he died and only attended the funeral.

Dying is one of the most frightening, lonely experiences a human being will go through, and we will all go through it. I have recently heard stories about people crawling into bed to sleep with and hold their parents while they were dying, at the request of the parent. While this may not be possible or appropriate in many situations, nurturing human touch can offer comfort, and soothe the person who is transitioning to death.

The need to belong, and to know you're part of the clan, is a powerful yearning. This need increases exponentially when life fixes you a nice shit sandwich. When you have experienced loss or heartbreak, nurturing human touch can make a huge difference in your healing process.

Two summers ago, my friend Justin was traveling in Europe,

counting down the days until their partner could join them. They got a call that their partner had been killed in a motor-cycle accident, and cancelled their trip in order to return home. When they arrived at their house, their friends had assembled with food, music, and dozens of hugs. They attribute this home-coming to easing their grief as they rebuilt their shattered life.

When you lose your partner, parent, pal, or pet, you can alleviate your suffering by asking for nurturing human touch. Bring together a group to support and love you through plac-ing their hands on your body. The memories of their love and caregiving through touch will sustain you long after people have stopped bringing by meals and sending sympathy cards.

All these suggestions are wonderful when you have a strong circle of friends, but how *do* you get your touch needs met when you are living in a new town and not surrounded by people you know? Or when you don't have (m)any friends? Don't worry, I've got you covered for that too. Read on!

CHAPTER 16

Touch with Strangers

WHILE OXYTOCIN IS often called the trust molecule, the love molecule, or the moral molecule, it has a darker side. It has been found to promote in-group bias: People are willing to choose someone from their own tribe at the expense of someone from a different group if given the choice.[1]

This is a feature, not a bug. When you think about it, it makes perfect sense: If oxytocin is internally generated and not snorted in a lab, it's coming from touch, connection, and intimacy – things you share with those who are close to you. Its greatest time of production is during childbirth, and right after childbirth when your baby is in infancy.

You want to protect your baby from harm, right?

People often debate whether or not nurturing human touch from strangers is going to promote feelings of trust and offer comfort. Many people shame those who seek out nurturing human touch from professionals, saying they are pathetic, or proclaiming that no male on the planet could possibly cuddle a woman without trying to have sex with them. If you have the opportunity to touch and be touched by people you love and trust, it should make you feel good. But there is something missing in this debate:

For many people, the choice is not between being touched

by someone you know and love, and being touched by a stranger. The choice is between being touched by a stranger and not being touched at all.

As you know your touch type better than anyone else, you are the only person who can decide if touch from a stranger will work for you. I can say, though, after holding hundreds of people, most of them strangers, that consensual touch with strangers is a vast improvement over remaining physically isolated and not being touched. The one word I use to describe people after they have their session is "relieved."

Don't Touch Strangers!

The fear that strangers are dangerous and want to harm you via touch – especially predatory men putting their hands on vulnerable women and children – is deeply ingrained in our psyche and culture. This has been thrown out a bit with hookup culture where people share sexual touch with near-strangers all the time, but in large part, we continue to perpetuate this narrative. Just as touching your friends consciously and carefully can be beneficial, we can do the same with strangers (same basic concept, slightly different script).

You might want to seek out touch with strangers if:

* You have recently moved to a new city (or part of the world) and don't have a network of people who can provide nurturing human touch.
* You feel more comfortable right now with strangers than friends (this is surprisingly common).
* You have been through a trauma, illness, or loss and feel unsupported, misunderstood, and/or don't want to burden your friends further.
* The idea of touching someone you already know feels too

intimate and risky.

* You are spending a lot of time being a caregiver of a new child or a sick partner and need someone to take care of *you.*

* Your family of origin didn't touch you much and/or you have little experience with intimacy and you want to practice and learn more.

* You have been single for a long time and can't or don't want to get your touch needs met through casual sex.

If you are single but don't want to be, nurturing human touch can make you more attractive to prospective partners. Touch hunger can make you come across as needy: Even if you aren't overtly grabbing people, others can smell eau d'esperation wafting out of your pores. Think about it: How would you act at a fancy dinner party with a seven-course meal if you arrived after not having eaten for three days? Your hosts would probably show you the door if you started shoveling the carpaccio down your throat like you were trying to make yourself into human foie gras.

Below you will find some of the various alternatives for getting your nurturing human touch needs met by strangers, and some tips and tricks that might make it go more smoothly.

Personals Ad for a Cuddle Buddy

I've been pleased to see a rise in the number of such ads coming through in my alerts, and have read plenty of them. This is a good method to find someone you can share touch with on a regular basis, and to ascertain that the person you are approaching is looking for the same thing you are.

If you want to run an ad like this, you should:

* Get clear on what you want. If you're advertising for cud-

dling, but really looking to hook up, don't bother. Most people can smell a bait-and-switch scheme from miles away. Leave phrases like "If it leads to more, I'm open to it" out of your ad. It might happen that way organically, but if you want nurturing human touch, write your ad accordingly.

＊ Make your prospective respondents feels safe. Be willing to meet in a public place initially, like a park or a cafe. When you do meet, ask your potential partners what kind of touch they want, and wait for them to answer. Find mutually agreed-upon activities instead of rolling with your own agenda.

＊ Be open-minded about appearance. To crib a line from *Pulp Fiction*, what's pleasing to the eye and pleasing to the touch are not always the same thing. Instead, look for shared interests so you have something to talk about while you're snuggling.

＊ Have realistic expectations. While you might end up creating a relationship with nurturing human touch that includes sleepovers, it's highly unlikely that you will find someone who wants to come over and spend the night from the get-go.

Free Hugs

It's not uncommon these days to see people holding up signs at festivals, or in busy urban areas, offering free hugs. Perhaps you want to challenge yourself to connect with people and you want a low-stakes way of doing so. Not only will you get a lot of hugs, you'll get plenty of opportunities to find out how people hug and you will probably make someone's crappy day much better.

If you want to offer free hugs, you should:

* Make a sign and stand in a spot with high foot traffic. Holler out your intentions if you want.
* Wait for people to approach you.
* Let the other person determine the length of the hug.
* Move if you are asked to do so, especially if you're on private property. (You'd be surprised how threatening human kindness is to large corporations.)
* Get a group of friends together and go out in a group, if possible.

ICU Baby Cuddling

Most hospitals around the world have teams of volunteers whose sole task is to hold and massage babies born prematurely. (The health benefits of this practice are a huge inspiration for Karuna Sessions.) And while you won't get the benefit of sharing touch with someone who can reciprocate or negotiate with you, you will be giving a new member of the human race a better start.

If you want to cuddle premature infants, you should:

* Contact local hospitals and find out what their need is.
* Know that you will have to submit to a thorough training and possibly a background check.
* Make a commitment to volunteer for a certain number of hours weekly.

Cuddle Parties

Since 2004, cuddle parties have allowed groups of strangers to come together and experience platonic touch. Some areas have robust communities of cuddling enthusiasts who get together on a monthly basis, while other places have them sporadically. They can be weird and awkward at first, but by the end of the evening, strangers have become friends, and smiles and tears are happening.

If you want to attend a cuddle party, you should:

* Google "Cuddle Parties" to see if you can find an event in your area. You can also check Meetup or your local weekly paper.
* Read as much about them as you can – there is tons of great information online.
* If you have a romantic partner, make sure they know what you're doing, and that you have their blessing. You could even invite them to come with you to check it out.
* Go with an open mind and be willing to say no.
* If there are no regular cuddle parties in your area, check with Monique Darling, an excellent facilitator who travels the country, to see if she has events planned in the area.

Professional Cuddlers

Professional cuddling is a rapidly growing industry. Some cities have a brick-and-mortar space where you can go and see a professional cuddler, while other areas have people who connect through websites and will either come to you or have you come to them. There are several different companies out there who offer this service.

If you want to see a professional cuddler, you should:

* Have an in-person or phone interview with them prior to your appointment.
* Make sure they have a contract and/or code of conduct.
* Figure out how payment works and plan accordingly.
* Shower and brush your teeth prior to your appointment.
* Wear comfortable clothes.

✴ Let a friend know where you are going (if possible) and text them before and after your appointment.

✴ Allow the cuddler to guide you through your session with questions and/or suggestions; they will have much more experience than you.

✴ Ask for a different type of touch or cuddle if there is something else you would like. You are paying for the session and it is your time.

Karuna Sessions

While at first glance Karuna Sessions appears to offer cuddling with two practitioners, it's much, much more. We have designed an experience of human connection that starts with an initiatory ritual and light touch, peaks with you being held in between two practitioners in a way that mimics the feeling of being held by your mother as an infant or surrounded by your tribe in prehistoric times, and ends with a tea ceremony. It's integrative medicine that treats body, mind, heart, and soul and leaves you with a plethora of data to contemplate and process long after your session is over.

If you want to have a Karuna Session, you should:

✴ Be able to come to Austin, Texas (alternatively, if we have enough appointments lined up in one location, we can travel).

✴ Put us in touch with your therapist if you have one so that we can work with them.

✴ Ask plenty of questions ahead of time.

✴ Follow the instructions on the door.

✴ Give yourself time to integrate your experience after.

Escorts

If you have read literature or research on sex work, you know that escorts commonly have clients who simply want to be held. While it's unlikely that they will offer you a discount because you are not seeking sex, seeing an escort for nurturing human touch is a practical option. It might also be your only option if you live in an area that doesn't have nurturing human touch professionals.

If you want to see an escort, you should:

* Request a phone conversation ahead of time to make sure they are a good match for you. Explain to them what you want and give them the opportunity to say no – it may feel too intimate for some sex workers.
* Be prepared to spend a good amount of money. (Note: It's usually a cash business.)
* Figure out where you will meet. Escorts who do in-call sessions will have you come to them, while out-call workers will come to your location.
* Let someone know where you are going (if possible) and text them before and after your session.

## CHAPTER 17

Don't Touch Me!

RECENTLY STUMBLED UPON a company that makes #AskMe t-shirts. They are designed to be conversation starters, and offer people an opportunity to talk about political, social, or personal matters with the person wearing them. I think my favorite thus far is #AskMe about a teacher who changed my life – I'm happy to tell you about my journalism professor Jake Highton who taught me how to write fast, and well, in my sophomore-year reporting class.

When I saw the shirts, my first thought was "Do they have t-shirts that say #AskMe For A Hug"? My second thought was "If not, will they make me one?" Not only did they have #AskMe For A Hug shirts, they also had #AskMe Before You Touch Me shirts. My friend Riley, who hates being touched, said they were going to order seven of 'em, one for every day of the week, and make it their costume the next time they go to a festival.

If you're touch-sensitive or touch-averse, you probably want seven of these shirts for yourself to wear to work, to the grocery store, to events, to the doctor's office, to family gatherings, and to bed at night. You will buy a crate of them, and dye them different colors to match each of your outfits. If you have your druthers, you will be buried in one of them so that people don't touch your corpse.

As they say in HashtagLandia, #thestruggleisreal.

This book was written for people who want more touch in

their lives, but as we have seen, touch is an individual need and desire that will probably change over your lifetime. This chapter is for people whose idea of getting their touch needs met is having much less touch in their lives.

Own Your Boundaries

After you graduate from Miss Epiphany's School of Nurturing Human Touch, you will discover that you have become much more *discerning*. Things you might have gone along with, or put up with, won't feel good anymore. Even if you are a touch enthusiast, there will be times when you don't want to be touched. There will be people you don't want to have touch you. There will be situations where touch feels inappropriate.

Navigating the boundaries between yourself and others is an ongoing task. Boundaries are fluid, and most people don't understand them. Ultimately, there is only one person whose approval you need to set boundaries around touch: you. This isn't always easy.

While our culture is one of touch deprivation, many subcultures are touchy-feely in a non-consensual way – hugging is accepted and expected. If you are touch-sensitive or touch-averse, you will find yourself in the minority in many social situations. One of the keys to being successful in defending your boundaries is accepting that they are uniquely yours to define, being confident in your choices, and being comfortable with establishing them regardless of what anyone else might think.

For many years, I have been fascinated with watching how people with habits/positions outside of the mainstream handle themselves when faced with setting boundaries. This past Thanksgiving, we had one person at our meal who ate only raw foods. They graciously brought along their own food, contributed to our group meal, and didn't criticize the rest of us for consuming animal products when we were rolling around on the floor in an overstuffed tryptophan haze. This is in contrast to another vegetarian friend of mine who goes on long rants about how offended they are when someone invites them to dinner and assumes that they can just eat side dishes. They have been a vegetarian for 20+ years but still feel rejected because of the choice they have made.

Being firm about your boundaries doesn't guarantee that you won't get pushback, but it can reduce the amount of friction you get as you move through the world. The fewer fucks you give about whether or not people like or respect you for having boundaries, the better you will fare. It's not your job to mollify them.

Don't Take It Personally

While alternative communities rebel against mainstream thinking and habits, there is a surprising amount of cultural conformity within any group. This is evident, and prevalent, when it comes to touch. Don't be surprised if you are met with defensiveness or anger from others when you say no to being touched.

Some people are attracted to alternative communities because they are interested in the activities or philosophies said groups espouse, but many people come to them because they are seeking to belong. When you have gone through life as an outcast, it is a huge relief to feel like you have found people who "get you." Fitting in, and instantly having a big social circle, can be intoxicating.

Once you have found your tribe, you discover that you cannot spend nearly as much time with them as you like. Those pesky responsibilities – work, school, and family – leave you with the occasional night or weekend to commune with your fellow fanatics. When you are together, you want to touch them, a lot, to make sure they are real, and that the bonds you have created are genuine. One of the most common ways this happens is by hugging people.

When you tell someone not to touch you or hug you, they may feel like you are rejecting them. They may also feel like you are rejecting their philosophies, their interests, and their community. This might not go over well if they are used to getting their way, or feel entitled to hug anyone they want to hug. They may guilt-trip you, or get angry.

Stick to your guns. If you don't want to be touched, you shouldn't do it to please others. Don't be afraid to go against the grain. I can guarantee that there are others in your group who wish they could do the same. They may feel more empowered if they see you doing it.

Worried that people will gossip about you? Don't be. They are going to gossip about you whether you set boundaries around touch or not.

Just Say No

The word "no" is a complete sentence. You are not required to give a further explanation if you don't feel like it, especially if you know the person who is pressing you for details can be argumentative or manipulative. Saying more than "no" may give them an opportunity to try to rationalize or wear down your boundaries. You can say "thank you, but no" if you want to be polite. Repeat as often as necessary.

Make a Plan

A few years back, I was on a diet that required me to eat every two to four hours. I quickly discovered that if I didn't stash some cheese sticks and almonds in my bag when I was out for the day, my choices would be a nasty case of psychoglycemia or snarfing down some non-compliant food I had scrounged up in a corner store.

If you know you will be in a touch-heavy environment and you will need to set boundaries, be prepared ahead of time. There are a few things you can do that will make this easier:

* Say it. When someone goes in for the hug, step back and say, "I'm not a hugger," "Thank you, but no," "I don't feel comfortable with hugging people I don't know well," or "I don't like to be touched." If you want to say "I'm just getting over a cold and might still be contagious," go for it, but people might get suspicious when your cold lasts for two years.
* Wear it. If you aren't comfortable with verbally setting boundaries, a shirt or button that says "Ask Before You Touch" can do most of the heavy lifting for you. People might not always see it, but you can always point to it. It will also give you opportunities to have conversations with people about why you don't like to be touched, or why it's important to ask for consent before you touch anyone. With a bit of luck, people will remember it the next time they encounter you.
* Be proactive. Before they go in for the hug, stick your hand out for a handshake, or hold your hand up like a crossing-guard in the universal body language for "STOP." You can also put your hands up in a prayer position and bow to them.
* Lean on your friends. It's helpful to have people who know about your discomfort with touch who will be willing to keep an eye out for you. These are the people you trust to touch you who will happily run interference, either by putting their arm around you when someone approaches

or jumping in front of you to take the hug bullet. With a bit of luck, your friends will also be willing to back you up verbally if/when someone takes offense at your lack of desire for physical contact. Sadly, sometimes it helps to hear it from somebody else.

* Stay clear. If your community has a penchant for celebrating with intoxicants, go to events early in the evening before people get wasted. Alcohol has a way of making people less inhibited, more open, and more inclined to ignore body language. Attending while people are still (relatively) clear-headed will let you see the people you like without having them slurring "I love you!" and trying to give you a big, sloppy hug.

* Make an exit plan. Know how long you will stay in a given situation. Stand close to the exit, or know where the exits are. Don't allow yourself to get guilted into staying. If you are with friends, make sure they know your desires before you go someplace, or arrange for your own way to get home if need be.

* Play with modeling consent. Perception and reality are two different things. While I believe you should ask before you touch someone, I won't ask for explicit consent from someone I've known for a long time – the consent has been given many years back. If someone is new to a group, though, they don't know I've had a long relationship with the person I'm hugging: What they see is that I just touched someone without consent. If I ask my long-time friend if they would like a hug to model good boundaries, a little light bulb may go on above others' heads that they should ask as well.

* Make an announcement. You can let everybody know about your preferences around touch on an email list, discussion group, or social media, but doing so might get you attacked. If you have spent any time online, you already know that

people tend to be much less empathetic with a screen to protect them, and often say all kinds of nasty things they wouldn't say to your face. Or they may not remember when they are "in the moment" and you will have to assert yourself all over again.

Own Your Desires

While this book is about non-sexual touch, I felt it important to include this section, as touch *is* a huge part of sexual relationships.

Whether you are touch-averse or a touch enthusiast, you have a responsibility to disclose your preferences and boundaries to potential partners when you are dating. Because we do expect our romantic partners to fulfill all of our touch needs, a relationship between two people with wildly different needs can be uncomfortable, disappointing, and hurtful. If you've ever been in a monogamous relationship where your sex drive was lower, or higher, than your partner's, you know what I'm talking about.

While you don't need to disclose this information on a first date, it is good to discuss before things get hot and heavy. Most of us will constantly touch our partners when we are first in a relationship. The beginning of a relationship where love and attraction is blooming includes lowered inhibitions and behaviors that may not be sustained once you settle into a routine.

If you are honest about what you want and need, you are more likely to find someone who wants and needs the same things you do. This will probably reduce your pool of potential partners. And while it may not be romantic, t's highly preferable to leaving somebody wondering what is wrong with them because they think you are rejecting them.

Boundary Crossers

While many people will cross your boundaries out of ignorance or carelessness, others do it deliberately, for their own amusement or gain. It is unlikely that you will make it through your life without encountering them. Below are a few things you can do to keep the discomfort they might cause to a minimum.

Know What to Look For

Many boundary-crossers are master manipulators. They are charismatic and likeable, and can be found in positions of power in workplaces and social groups. Oftentimes they work hard to contribute to their organizations so that they can become irreplaceable.

One of the best ways to avoid becoming a victim is to assert your boundaries from the get-go. Moving away from a wandering hand, looking someone in the eye, and saying "Don't touch me" the first time they cross a boundary lets them know that you are not going to be passive about their transgressions. This can also extend to shutting down inappropriate comments as soon as they are made, especially in the workplace. Boundary crossers will test you for weaknesses, so don't show them any.

Get a Second Opinion

When you are in a new social circle or work situation and someone makes you uneasy, ask around. If somebody is a serial boundary-crosser who has not been asked to leave the group, their reputation will be well-known. The only recourse is for the members with less power to protect themselves and others through whispered advice. When people have been unable to eject a boundary crosser from an organization or social group they usually do their best to make sure no one else is hurt by that person.

Speak Up

You can try to publicly denounce a boundary crosser's actions, but you will probably get pushback. If something happened in a private setting without witnesses, it will be one person's version of events against the other person's versions of events, making it hard to prove. In addition, many people still shrug off boundary crossing as "no big deal" – you may not get the support you had hoped for, especially if the boundary crossing doesn't fit the outsiders' definition of assault.

It is a human trait to take the path of least resistance, and most people will not speak up when they see wrongdoing, nor can you expect them to have your back. Consider your own needs first, especially if the boundary-crossing incident has left you traumatized. It's not easy to be fighting against not only a predator, but a system that actively protects predators, while you are in a vulnerable position.

Get Out!

If you find yourself in a situation with someone who will not take no for an answer, or with someone whose prior boundary-crossing escapades have been ignored, your only recourse may be to extricate yourself from your workplace or social circle. While this can be frustrating, heartbreaking, expensive, or annoying, it's better than continuing to expose yourself to risk, or watch others get hurt.

Conclusion

MY FIRST ENCOUNTER with an alternative community was a Grateful Dead show in 1982. My bestie Mike, and I, always desperate for live music in Reno, had bought tickets for a band we knew little about. When we showed up at the Centennial Coliseum, we were shocked to see hordes of tie-dye-adorned, patchouli-reeking people who had driven from around the country to worship at the altar of Jerry Garcia. They were friendly and welcoming, and handed out roses and hugs to their fellow Deadheads.

The Deadheads' colorful wardrobe and love for their favorite band and each other vivified the pale gray concrete stairs and blue plastic seats of the Coliseum. It also revealed that there was a possibility of connection and community for me after a childhood as an outsider in a small, conservative city.

I attended about 20 Dead shows after I moved to San Francisco in the 80s, but wasn't crazy about the music. I made a few friends, but never felt like I belonged. Even though they were not the 'droids I was looking for, I appreciated the way people watched out for each other. While I had several friends who attended the Dead 2.0 reboot the past couple years, I never considered going to see them.

"Never say never" is both a tired cliche and a challenge to the universe. On an unseasonably warm December afternoon, I found myself across the street from a Dead and Company show with my friend Lauren who had come to Austin from Salt Lake City. The faithful had set up Shakedown Street, an open-air bazaar, on a triangle of grass next to I-35, and were selling posters, stickers, t-shirts, and grilled cheese sandwiches. Several vendors had brought nitrous tanks, and the ground was littered with glassy-eyed hippies and empty balloons. We wandered through the crowd looking for gifts for Lauren's friends...and hugging people.

I had suggested to Lauren that we whip up "Free Hugs" signs and distribute oxytocin to the faithful, and she readily agreed. People's faces lit up when they saw our homemade signs. We hugged dozens of people in an hour. Some of them barely touched me; others melted into my arms for a minute-long hug. I got side hugs, and bear hugs. Several people held out their arms and grabbed Lauren and I for a triple hug. One woman jumped off a low wall, ran at me, squealed "free hugs!" and threw herself into my arms.

Nobody mistook our affection for a sexual come-on.

Lauren, who is warm-hearted and makes friends easily, was amazed. "People are starving for connection," she said. "This is crazy!" Weeks later, she was still talking about the experience.

Thousands of people had come to bond with others over the music, in a scene that was often referred to as a family. But even here, people felt isolated and alone, and were thankful for a hug from a stranger. This brief physical contact made them smile and provided them with tangible evidence that they belonged.

Loneliness Is An Epidemic

If people struggle with finding connection in a group known for its openness, what of people who don't have a community? There are billions of people living alone in small spaces in big cities. They may be surrounded by other people at their workplaces, and on the street, but they are sequestered in their heads and their bodies, a community of one.

After five years of contemplating nurturing human touch,

I can see in somebody's face when they haven't been touched tenderly in a long time. It's as obvious to me as a broken bone.

The World Health Organization predicts that there will be two billion people over the age of 60 by 2050, and many of them will have their lives and health negatively impacted by loneliness. People in high-income countries like the United States are at greater risk. Gone are the days when elders were an integral part of their tribes, valued for their wisdom and experience.

But loneliness has already become an epidemic, and it affects people of every age. When I look around, I see people who are hurting and need comfort. They might try to mask it with a brusque "I'm great!" when asked how they are doing, but dig a bit deeper and you hear how fucking tired everyone is. Even those who seem to "have it all" with great salaries, big houses, decent health insurance, and domestic help are over-scheduled and tired of trying to keep their nuclear-family-size clans safe and secure.

It gets worse as you go down the ladder. What's going on with those who are living paycheck to paycheck? The elderly folks working at Wal-Mart? The single moms trying to raise kids on a server's salary? The guys standing in front of Home Depot desperately trying to pick up some day labor? They struggle mightily with keeping their tribes afloat, especially if their tribe only contains one person.

The Trimtab of Touch

I have many smart, thoughtful friends who do their damnedest to pitch in and stay informed. Solving our collective problems is a complex task. It's exhausting to figure out all the moving parts and take everybody's needs into account. Most people I know are overwhelmed at the prospect of changing the world.

Nurturing human touch is a simple and immediate way to

make the world a better place. It doesn't need to be legislated or mandated. It doesn't require a huge team of people with vast financial resources. It can be effective without a ton of planning, or special facilities and equipment. It *will* require slowing down, seeing people, and recognizing that we need to take care of each other. Because if we don't start right where we are, right in this moment, what else can we do? There is no easier, quicker way to make someone's world a little better than kindness delivered via consensual, nurturing human touch. Sometimes, it can change a person's life.

Buckminster Fuller used the metaphor of trimtabs to describe how an individual can create social transformation. ("Call me trimtab" is on his gravestone!) A trimtab is a small piece on the edge of an airplane wing or boat rudder that moves into the currents that oppose it, and leverages that power. While the trimtab itself is rigid, it is on hinges so it can be moved around. A trimtab gradually changes the trajectory; a small course correction becomes wider as the vessel moves forward.

I am not naive enough to believe that nurturing human touch will solve all of our problems, but I believe it can serve as a trimtab for humanity. I believe that it can provide us with the feeling of belonging and connectedness that many of us are missing. The world is changing rapidly; we are making leaps and bounds with technology, but have yet to evolve our hearts and consciences at the same speed. Our bodies are being ignored in favor of our minds. Isn't it time we gave our bodies the connectedness technology has provided our psyches?

Helping Humanity Survive and Thrive

Our species survived by remaining physically close to one another. The youngest members of our tribe thrive when they are held and cared for by their parents, and the effects of this early loving touch resonate long after they are out of diapers. Call me crazy, but maybe it really is that simple.

The world I want to live in recognizes the value of nurturing human touch. I envision a world where people are cared for, seen, and held. A world where people turn toward each other, instead of away. A world where we are no longer ashamed of having bodies, and we recognize that our minds work better when our bodies are treated kindly. A world where people have the agency to ask for what they want and need, and their no's are respected. A world where we can play and laugh and enjoy the company of our fellow humans, if only for a moment.

Let's start, like, now. The choice is yours.

Would *you* like a hug?

Endnotes

Chapter 3

1. Jourard, S.M., Disclosing Man to Himself, Van Nostrand Reinhold Company, 1968.
2. Wittig, R.M., et al., Food sharing is linked to urinary oxytocin levels and bonding in related and unrelated chimpanzees, Proceedings of the Royal Society, Volume 281, Issue 1778, March 7, 2014.
3. Vespa, J., et al., America's Families and Living Arrangements: 2012 Population Characteristics (Aug. 2013) US Census Bureau.
4. CDC/NCHS, National Vital Statistics System.
5. US Census Bureau 2010.
6. Holt-Lunstad R., et al., Loneliness and Social Isolation as Risk Factors for Mortality: A Meta-Analytic Review. *Perspectives on Psychological Science*. 2015.
7. Bureau of Labor Statistics, American Time Use Summary Table 11A, Time spent in leisure and sports activities for the civilian population, by selected characteristics, averages per day, 2017 averages.
8. Armstrong, Martin, Smartphone Addiction tightens its global grip, Statista, May 24, 2017.
9. McPherson, M., et al., Social Isolation in America: Changes in Core Discussion Networks over Two Decades, *American Sociological Review*, Vol 71, Issue 3 (2006).
10. Bohn, R.E., How Much Information? 2009 Report on American Consumers, *ResearchGate*, January 2009.
11. Twenge, J., Have Smart Phones Destroyed a Generation? *The Atlantic*, September 2017.
12. Abma, J.C., Sexual Activity and Contraceptive Use Among Teenagers in the United States, 2011-2015, National Health Statistics Reports, Number 104, June 22, 2017.
13. Ahrnsbrak, R., et al., Key Substance Use and Mental Health Indicators in the United States: Results from the 2016 National Survey on Drug Use and Health.
14. Harvard Medical School, National Comorbidity Survey, August 21, 2017.
15. Florence C.S., Zhou C., Luo F., Xu L., The Economic Burden of Prescription Opioid Overdose, Abuse, and Dependence in the United States, 2013. *Med Care*. 2016;54(10):901-906.
16. Statistics gathered by the American Pet Products Association.

18. Nagasawa, M., et al., Oxytocin-gaze positive loop and the co-evolution of human-dog bonds, Science, Vol. 348, Issue 6232, April 17, 2015.
19. US Census Bureau, America's Families and Living Arrangements: 2016, Table A-1.
20. Cohn, D., et al., Barely Half of US Adults are Married - A Record Low, Pew Research Center Social and Demographic Trends, 2011.
21. Billings, J.D., Hardtack and Coffee or the Unwritten Story of Army Life, Albion Press, 2015.

Chapter 4

1. Muroyama, Y., Mutual reciprocity of grooming in female Japanese macaques (Macaca fuscata). (1991) *Behaviour* 119, 161-170.
2. Ferber, S.G., et al., The Effect of Skin-to-Skin Contact (Kangaroo Care) Shortly After Birth on the Neurobehavioral Responses of the Term Newborn: A Randomized, Controlled Trial. (2004) *Pediatrics* 113(4), 858.
3. Field, T., Massage Therapy for Infants and Children. (1995) *Journal of Developmental and Behavioral Pediatrics*, 16, 105-111.
4. Harlow, H.F., Love in Infant Monkeys. (1959) *Scientific American* 200, 68-74.
5. Scafidi, F.A., et al., Massage stimulates growth in preterm infants: A replication. (1990) *Infant Behavior and Development,* 13, 167-188.
6. Wheeden, A., et al., Massage effects on cocaine-exposed preterm neonates. (1993) *Journal of Developmental and Behavioral Pediatrics*, 14, 318-322.
7. Hertenstein, M.J. et al., Emotion Regulation Via Maternal Touch. *Infancy.* (2001), 2(4), 549–566.
8. Wright, B., et al., Evidence-Based Parenting Interventions to Promote Secure Attachment: Findings From a Systematic Review and Meta-Analysis (2016) *Global Pediatric Health*, Vol. 3, 1-14.
9. Feldman, R., et al., Maternal-Preterm Skin-to-Skin Contact Enhances Child Physiologic Organization and Cognitive Control Across the First 10 Years of Life. (Jan. 2014) *Biological Psychiatry*, 75:1, 56-64.
10. Hertenstein, M., et al., Touch Communicates Distinct Emotions, (Aug. 2006) *Emotion* 6(3), 528-533.
11. Maksimovic, Srdjan, et al., Epidermal Merkel cells are

mechanosensory cells that tune mammalian touch receptors, (May 2014) *Nature* 509, 617-621.

12. Cauna, N., Nature and functions of the papillary ridges of the digital skin, (Aug. 1954) *The Anatomical Record*, 119:4, p. 449-468.

13. Light, K.C., et al., More frequent partner hugs and higher oxytocin levels are linked to lower blood pressure and heart rate in pre-menopausal women, (Apr. 2005) *Biological Psychology*, 69:1, p. 5-21.

14. Moghimian, M. Ph.D., et al., The role of central oxytocin in stress-induced cardioprotection in ischemic-reperfused heart model, (Jan. 2013) *Journal of Cardiology*, 61:1, 79-86.

15. Elabd, C., et al., Oxytocin is an age-specific circulating hormone that is necessary for muscle maintenance and regeneration, (June 2014) *Nature Communications*, 5, p. 1-11.

16. Morhenn, Vera, et al., Massage Increases Oxytocin and Reduces Adrenocorticotropin Hormone in Humans, (Nov. 2012), *Alternative Therapies in Health and Medicine*, 11-19.

17. Poutahidis, T., et al., Microbial Symbionts Accelerate Wound Healing via the Neuropeptide Hormone Oxytocin, (Oct. 2013) *PLoS One*, 8(10), 1-17.

18. Nation, D.A., et al., Oxytocin Attenuates Atherosclerosis and Adipose Tissue Inflammation in Socially Isolated APoE-/- Mice, (May 2010) Psychosom Med. 72(4); 376-382.

19. Kovacs, G.L., et al., Oxytocin and Addiction: a Review, (Nov. 1998) *Psychoneuroendocrinology*, 23:8, 945-962.

20. Bowen, M.T., et al., Oxytocin prevents ethanol actions at subunit-containing $GABA_A$ receptors and attenuates ethanol-induced motor impairment in rats, (2015), *PNAS*, 112:10, 3104-3109.

21. Blevins, J.E., et al., Role of oxytocin signaling in the regulation of body weight, (Dec. 2013) *Rev. Endocrin. Metab. Disord.*, 14(4), 311-329.

22. Pavlov, V.A., et al., The vagus nerve and the inflammatory reflex - linking immunity and metabolism. *National Review of Endocrinology*, (Dec. 2012) 8:12, 743-754.

Chapter 6

1. Office of the Administration for Children and Families, Child Maltreatment Report, 2011, p. 21.

2. Kilpatrick, D.G., Youth Victimization: Prevalence and Implications, National Institute of Justice Report, 2003, p. 5.

3. Black, M.C., et al., The National Intimate Partner and Sexual Violence Survey: 2010 Summary Report.

Chapter 10

1. Burkeman, O., This column will change your life: Are you an Asker or a Guesser? *The Guardian*, May 7, 2010.

Chapter 16

1. De Dreau, C.K.W., et al., Oxytocin promotes human ethnocentrism, *PNAS*, Vol. 108, Issue 4 (January 2011).

Resource Guide

This list is not exhaustive; it has been curated for quality over quantity.

For the Mind

Touching: The Human Significance of the Skin by Ashley Montague

Touch: The Science of the Hand, Heart and Mind by David J. Linden

Touch by Tiffany Field

Ask: Building Consent Culture edited by Kitty Stryker

Pagan Consent Culture edited by Christine Hoff Kraemer and Yvonne Aburrow

The Consent Guidebook: A Practical Approach to Consensual, Respectful, and Enthusiastic Interactions by Erin Tillman

The Snuggle Party Guidebook: Create Deeper Friendships, Decrease Loneliness, & Enjoy Nurturing Touch Community by Dave Wheitner

For the Body

Cuddlist Professional Cuddlers
https://cuddlist.com/

The Snuggle Buddies
http://snugglebuddies.com/

Touch Is Medicine
https://www.touchismedicine.com/

Cuddle Sanctuary
https://cuddlesanctuary.com/

Cuddle Up to Me
https://cuddleuptome.com/

Holding Space
https://www.holdingspacellc.com/

Cuddle Party
http://www.cuddleparty.com/

Betty Martin
https://bettymartin.org/

Monique Darling
https://whereintheworldismoniquedarling.com/

The Snuggery
http://www.thesnuggery.org/

Karuna Sessions
http://www.karunasessions.com

CuddleOM
https://www.cuddleom.com/

Cuddle Professionals International
https://www.cuddle-professionals.co.uk/

Snuggle with Sam
https://www.snugglewithsam.com/

Acknowledgments

When I think about an author writing a book, the image that comes to mind is of a person sequestered in a tall tower, a tiny room, or a cabin in the woods, toiling away in solitude, with nothing but the sound of the pen scraping against the paper or the occasional footsteps of the person dropping off meals to break their concentration. While my experience started out a bit like that – just substitute living room couch, laptop, and cat demanding breakfast– the truth is that putting a book together is a team effort.

I got thoughtful feedback on the first two drafts from my beta readers: Melissa Burnham, Benjamin Wachs, J. Absinthia Vermut, Kaia Tingley, Crystal Engelmann, Melody Byrd, Jill Marlene, and Heather Vescent. Thanks for being tender and enthusiastic about what you read and catching errors large and small. Your suggestions made the book much more readable and engaging. I also want to thank my advance reviewers who took time out of their busy schedules to read my words and then lend me their sterling reputations by saying nice things about what I wrote: Janet Trevino, James Fadiman, Halycon, Mo Daviau, Kim Corbin, Kitty Stryker, and Christine Hoff Kraemer.

I couldn't have asked for a better team of advisors and experts to walk me through the minutiae of publishing by adding their knowledge and savvy to the project. Melissa Kirk, my content editor, gave me suggestions both logical and intuitive, and walked me through my first-timer jitters. Lisa Pimental and her red pen tightened up my prose and made sure each litany of nouns included an Oxford comma. Mark Weiman gave me fantastic advice about book publishing, and created a beautifully-designed interior and exterior. And my longtime friend Eliza

Bundledee pulled my vision out of my head to create a vibrant, vulnerable cover illustration that perfectly captures the spirit of the book.

Without the crowdfunding platform Kickstarter, and the support of my formidable network, this book would exist only as a bunch of Google Docs and a dream in my mind. In the months prior to the Kickstarter launching, I had the brains, hearts and hands of Julie Gillis, Wendy Corn, Laura Bryan, Margot Duane, Kabi Nyoike, Michelle Seymour, M Bunny, Christopher Lucas, Dave Loomis, Jaguar Reow, Brackin Camp, and Alicia Prince contribute to creating and running a successful campaign. Only about one in three Kickstarters successfully funds, and I have little doubt mine would have been in the failure pile without the support of these fine folks.

My Kickstarter SuperBackers who believed in this project and put their money where their hearts are: Fei Wyatt, Obi Juan, Benoit Roche, Laurie Sargent Erney, Laura Gasparrini, Veek, Anonymous, John Nettle, Jonathan Kessler, Ani Colt, Tickled Pink TT, Beverly Marshall Saling, Goddess Gita, Sister Mable Syrup, Cristina Vincent, Jamie Okubo, Michael Prang, Lori Lakin Hutcherson, Meredith Theaker, LovelyLisaLou, Dan Shick, DanceSmiths Ballroom Dance Studio, Julie Alexander, Lisa Meece, Holding Space, LLC, Carl T, Susan E. Mazer, Ph.D., Kam, kerry kitrell art, Kevin Mathieu, Rhonda, Sherri Ferraro-Zaro, Roberto, Elaine Gerber, Johanna Berke, Evan Prodromou, Yoms, Eileen Piekarz, Anatole Barnstone, Claire James, Dana and Luis Ayala, Jane Brooks, Tricia Murphy, Mary Black, Crafty Avenger, Cabiria Baloney, Blair Miller, Steve Piasecki, Dena Crowder, Andrea Kaul, Gino & Kristie Essa, Amani Ellen Loutfy, Tammi Perry, Vanessa Filkins, Larry Edelstein, Bedpost Confessions, Suzanne Loosen, Tanya Sweeney, Sartaj Chowdhury, Shelley Imholte, Rhett & Amy Dawson, Tressa Piekarz, Kelsey Hitchingham, Laura LaGassa, Erik Johnson, Lisa Dugan

Manor, Jay Wu, Steve Fenoglio, and AnnMarie Sands. Thank you for believing in this work, and helping it get out into the world. Y'all rock..*and* you have the best names.

Many people told me that I made the acts of writing and fundraising look easy, but that was only because Julia Crenshaw-Smith and Bella LaVey did the hard emotional labor of letting me cry on their shoulders and listening to hours of me process my self-doubt and fears. I couldn't have wished for two better platonic trophy wives than these two. They have made my life richer, funnier, and softer for years.

Special shout-out going to Face, my favorite clown. Bet you didn't realize that you were the catalyst for this book, but I'm glad it happened that way. I miss our conversations and our hugs.

I could write an entire chapter about the appreciation and gratitude I have for my wonderful mother, Eva. She has unconditionally loved and supported me from weird, awkward kid crying alone on the playground through rebellious young adult making poor life choices to late bloomer who has found a way to make the world a better place. Thanks, Mom – dinner's on me next time we see each other.

And finally, thank YOU, dear reader, for taking a chance on something weird and new. I hope that working through the shame and vulnerability to reach for tangible, physical connection makes you a happier person, and shows you that we can find different ways to live and relate. The world could use an army of kindness about now, and you are its foot soldiers.

EMJ
Austin, Texas
August 2018

I have two favors to ask before you go:

1. The next time you see somebody having a bad day, ask them if they would like a hug. Or ask for a hug yourself if you're struggling. Kindness and caring FTW.

2. You know at least five people who need this book. It could be a coworker, a cousin or a friend. If you would recommend it to them, or gift them a copy, that would be swell. A nudge from someone they trust could lead them to new possibilities for a healthier, happier life.

Thank you!

www.nurturinghumantouch.com

CPSIA information can be obtained
at www.ICGtesting.com
Printed in the USA
FFHW020132121218